Securing Commercial Nucleic Acid Synthesis

FORREST W. CRAWFORD, KYLE WEBSTER, GERALD L. EPSTEIN,
DEREK ROBERTS, JOSEPH FAIR, SELLA NEVO

T0306860

RAND

For more information on this publication, visit **www.rand.org/t/RRA3329-1**.

Published by the RAND Corporation, Santa Monica, Calif.

Library of Congress Cataloging-in-Publication Data is available for this publication.

ISBN: 978-1-9774-1341-3

Cover: huenstructurebio.com/Adobe Stock.

About This Report

On-demand gene synthesis is a growing industry that has democratized access to customized synthetic DNA and RNA products used in biological research. This emerging technology may present risks that could allow nefarious actors to obtain sequences of dangerous organisms or novel-engineered pathogens to construct a biological weapon. Securing nucleic acid synthesis has emerged as a federal policy priority. The 2022 *National Biodefense Strategy and Implementation Plan for Countering Biological Threats, Enhancing Pandemic Preparedness, and Achieving Global Health Security* issued by the White House recommends enhancing screening of nucleic acid synthesis orders and developing systems for identifying and reporting suspicious orders. Executive Order 14110 tasks the White House Office of Science and Technology Policy with establishing a framework "to encourage providers of synthetic nucleic acid sequences to implement comprehensive, scalable, and verifiable synthetic nucleic acid procurement screening mechanisms, including standards and recommended incentives" (Executive Order 14110, 2023, p. 75201). Customers who receive federal funding will be required to obtain synthetic gene products only from providers who adhere to the new framework. The *Framework for Nucleic Acid Synthesis Screening* describes basic responsibilities of synthetic sequence providers.

In this report, we provide recommendations for securing commercial nucleic acid synthesis services against misuse. We reviewed U.S. government screening guidelines, the new government framework, synthetic biology industry consortium requirements, state legislative proposals, and the scientific literature to answer the following research questions:

- What is the current state of commercial gene synthesis screening?
- How should providers determine whether an order is suspicious?
- How can providers or third parties ensure that suspicious orders split across providers will be detected?
- How should information about suspicious orders be relayed to law enforcement?
- How should providers' conformity with screening rules be verified?

Implementation of the recommendations described in this report would satisfy the requirements of several components of Executive Order 14110, helping to secure the synthetic nucleic acid industry against its products being misused to create a dangerous biological weapon. The recommendations first propose an overall screening workflow for providers to implement, followed by recommendations to establish a database to log information about orders, implement customer screening, communicate with law enforcement, and assess providers' conformity with the screening framework.

The intended audience for this report consists of U.S. government officials and their agencies tasked with responding to Executive Order 14110 section 4.4(b), synthetic biology industry stakeholders, and synthetic nucleic acid customers.

Meselson Center

RAND Global and Emerging Risks is a division of RAND that delivers rigorous and objective public policy research on the most consequential challenges to civilization and global security. This work was undertaken by the division's Meselson Center, which is dedicated to reducing risks from biological threats and emerging technologies. The center combines policy research with technical research to provide policymakers with the information needed to prevent, prepare for, and mitigate large-scale catastrophes. For more information, contact meselson@rand.org.

Funding

Funding for this work was provided by gifts from RAND supporters.

Acknowledgments

We thank Allison Berke, James Diggans, Kevin Esvelt, Zeno Fox, Max Langenkamp, Michael Parker, and Sana Zakaria for helpful comments on the draft report. We are especially grateful to Tessa Alexanian for detailed feedback and advice on background material and recommendations. We thank RAND reviewers Alison Hottes and Shira Fischer for their constructive comments, which have substantially improved this report.

Summary

The increasing availability and decreasing cost of custom synthetic nucleic acids presents a risk of misuse. Made-to-order synthetic nucleic acids could be used to reconstruct an existing but hard-to-obtain pathogen or to engineer a novel or more-virulent pathogen, leading to the construction of a biological weapon. Securing nucleic acid synthesis is a policy priority for the U.S. government and the synthetic biology industry. Executive Order 14110 tasks the White House Office of Science and Technology Policy with establishing a framework "to encourage providers of synthetic nucleic acid sequences to implement comprehensive, scalable, and verifiable synthetic nucleic acid procurement screening mechanisms, including standards and recommended incentives" (Executive Order 14110, 2023, p. 75201). Executive Order 14110 further requires agencies to restrict federal funding only to entities that obtain nucleic acid synthesis services from providers that adhere to the screening framework.

There is currently no legal requirement that gene synthesis providers screen their customers or their orders to ensure that the requested synthetic DNA and RNA constructs are not used for harmful purposes, although many providers screen orders voluntarily. The International Gene Synthesis Consortium (IGSC), a consortium of industry gene synthesis companies, requires members to adhere to requirements in its Harmonized Screening Protocol. In 2010, Office of the Secretary of the U.S. Department of Health and Human Services (HHS) issued voluntary guidance on ordering and customer screening for providers of synthetic nucleic acids. This voluntary guidance was updated in 2023 by the Administration for Strategic Preparedness and Response, also within HHS. But there is currently no universally accepted standard for screening customers or orders. Not all companies screen their orders, and companies that do screen orders may follow different protocols or have different criteria for rejecting or reporting suspicious orders.

In this report, we review government screening guidelines, industry consortium requirements, legislative proposals, and the scientific literature to assess the current state of commercial gene synthesis screening; summarize proposals for strengthening screening; and contribute recommendations for implementing comprehensive screening policies to secure commercial nucleic acid synthesis. These recommendations expand on and strengthen government communications, such as the HHS guidance, industry consortium rules and best practices (such as those developed by IGSC) and proposals developed by researchers and provider organizations. Our recommendations also go beyond the *Framework for Nucleic Acid Synthesis Screening* (Fast Track Action Committee on Synthetic Nucleic Acid Procurement Screening, 2024) in describing specific steps and decision criteria in the screening workflow, along with the conformity assessment tools providers should use in designing screening procedures to comply with the framework.

The recommendations in this report address the following components of Executive Order 14110 (2023):

- 4.4(b)(i)(B) "determine standardized methodologies and tools for conducting and verifying the performance of sequence synthesis procurement screening, including customer screening approaches to support due diligence with respect to managing security risks posed by purchasers of biological sequences . . . and processes for the reporting of concerning activity to enforcement entities" (p. 75201)
- 4.4(b)(ii)(A and D) "develop and refine for possible use by synthetic nucleic acid sequence providers: (A) specifications for effective nucleic acid synthesis procurement screening; . . . and (D) conformity-assessment best practices and mechanisms" (p. 75202)
- 4.4(b)(iv)(A) "develop a framework to conduct structured evaluation and stress testing of nucleic acid synthesis procurement screening" (p. 75202).

The recommendations also complement proposals for international mechanisms to secure nucleic acid synthesis.

Key Findings

Our review of government nucleic acid synthesis screening guidelines, industry consortium requirements, legislative proposals, and the scientific literature yielded the following findings:

- Commercial gene synthesis presents a serious risk of misuse or accident.
- Screening protocols vary across providers.
- Providers currently have no way to share information about suspicious customers or orders.
- Strengthening customer screening and verification is a promising way to help secure gene synthesis.
- Clearer standards for reporting suspicious orders to authorities would streamline follow-up screening and improve enforcement.
- The U.S. government currently has no mechanism for assessing the conformity of providers' screening systems with guidelines or policies.

Recommendations

We recommend that the U.S. government take the following actions to address the components of Executive Order 14110 described above. The first recommendation proposes the overall screening workflow to be implemented by providers. Subsequent recommendations describe in more detail several steps in this workflow, including customer screening, checking a database for evidence of split orders, and communicating with law enforcement about suspicious orders. The final recommendation covers assessment of providers' conformity with the proposed screening framework.

- Revise the 2023 Administration for Strategic Preparedness and Response guidelines to include specific information to be collected and decision criteria for each step of the screening pipeline.
- Establish a third-party database to log information about synthetic nucleic acid orders and periodically check those records for evidence of suspicious activity.
- Establish customer screening requirements, including identity verification measures.
- Establish criteria and data requirements for sharing information about suspicious orders with law enforcement agencies.
- Implement procedures and standards for assessing the conformity of providers' screening practices with requirements.

Contents

Figure

CHAPTER 1

Background

Nucleic acid sequences are the physical representations of heritable biological information for all forms of life. In the forms of deoxyribonucleic acid (DNA) and ribonucleic acid (RNA), they define the genetic sequences of all plants, animals, bacteria, fungi, and viruses. Since the discovery of the structure and function of the DNA molecule in the 1950s, biologists have sought to characterize genetic information by sequencing the genomes of organisms. Knowledge about these sequences is now widespread, with biological sequences stored in such publicly available databases as GenBank. At the same time, scientists have invented tools for altering genetic information by inserting, deleting, or rearranging genomic elements in living systems. Progress in synthetic biology has advanced rapidly, providing laboratory protocols, machines, and reagents for rapid and accurate reading, writing, and editing of nucleic acid sequences (National Security Commission on Emerging Biotechnology, 2024a). Biotechnology has also advanced alongside new tools from synthetic biology, providing methods that allow researchers to construct customized or novel nucleic acid sequences and insert them into the genomes of existing agents or create novel agents. Synthesis of DNA and RNA molecules directly from digital sequence representations constitutes a major step forward in synthetic biology, connecting computational methods for designing genes to their physical manifestation for use in living organisms.

This chapter describes the current state of the commercial gene synthesis industry, the risk of misuse of synthesized nucleic acid sequences, related risks of benchtop synthesis hardware, and government and industry recommendations and requirements for synthetic nucleic acid screening. We also explain the parts of Executive Order 14110 (2023) that relate to securing nucleic acid synthesis and describe the scope and intended audience of this report. In this chapter and throughout the report, we use the following terms as defined by the U.S. Department of Health and Human Services's (HHS's) Administration for Strategic Preparedness and Response (ASPR):

- *Provider*: The entity that synthesizes and distributes double-stranded DNA (dsDNA). A Provider is understood to be an entity synthesizing dsDNA for and distributing dsDNA to a customer, not a research scientist collaborating with a colleague.
- *Customer*: "The individual or entity (such as an institution) that orders or requests synthetic nucleic acids from a Provider, or that purchases synthesis equipment from a Manufacturer."

- *Principal user*: "The individual who originates an order or request for synthetic nucleic acids or synthesizes nucleic acids and oversees the use of ordered or synthesized nucleic acids in the laboratory. The Principal User may also be the End User."
- *End user*: "The individual who possesses and uses synthetic nucleic acids that they have received from a Customer, Principal User, or another End User." (ASPR, 2023b, p. 3)

Commercial Gene Synthesis Market Grows as Costs Decrease

Commercial on-demand nucleic acid synthesis is a growing industry that has expanded access to customized synthetic DNA and RNA products used in biological research. Individuals and organizations can now order synthetic DNA or RNA from an online platform and have it delivered to their door by mail. A report from the National Academies of Sciences, Engineering, and Medicine (2018, pp. 18–19) describes nucleic acid synthesis as one of the key technologies enabling the synthetic biology revolution:

> The heart of synthetic biology is the ability to make DNA constructs quickly and efficiently. Improvements in synthesis technology have followed a "Moore's Law–like" curve for both reductions in costs and increases in the length of constructs that are attainable. These trends are likely to continue.

Minshull and Wagner (2009, p. 800) write of the promise of beneficial uses of nucleic acid synthesis services:

> Gene synthesis enables a new world of possibilities: in the development of biofuels to combat climate change, in drug development to combat both persistent and emerging diseases, in agriculture to engineer crops that are more nutritious and resilient, and in research to bring a deeper understanding of the inner workings of the cell and of life itself.

Commercial synthesis of short (approximately 13 to 25 nucleotides) single-stranded sequences of DNA—called *oligonucleotides*—has been available for many years, but oligonucleotides must be assembled using time-consuming laboratory techniques to piece together longer gene-length DNA sequences (Smith et al., 2003). Synthesis of much longer (ranging from hundreds to hundreds of thousands of base pairs) gene-length dsDNA is now feasible (Gibson et al., 2009). Providers offer gene-length products, up to 200,000 base pairs, in the form of nucleotide sequences in quantities ranging from a few micrograms to several grams. The commercial market for synthetic gene products has expanded, and costs have dropped sharply (Carlson, 2022). In addition to longer sequences, gene synthesis providers offer rapid and inexpensive synthesis and delivery of custom sequences. Current prices for oligonucleotides are as low as $0.07 per base pair and about $0.09 per base pair for gene-length sequences (Twist Bioscience, undated), and order fulfillment can occur in as few as two to three days.

Accessible Gene Synthesis Presents Risk of Misuse

Increasingly, accessible commercial gene synthesis reduces barriers to developing a dangerous pathogen (Langenkamp, 2023), including resurrecting an extinct pathogen, re-creating an existing pathogen that is rare or whose possession or use is restricted (e.g., select agent pathogens), or de novo engineering and synthesis of pathogens or novel biotoxins. Four major factors contribute to the risk of misuse: (1) ready availability of sequences for dangerous pathogens, (2) limited oversight of sequences for dangerous organisms by existing regulatory authorities, (3) the possibility of novel-engineered sequences that do not closely resemble those of regulated sequences, and (4) the market effects of declining costs for gene-length synthesis.

Pathogen Sequences Are Available Online

Synthesis of potentially dangerous extinct or existing but hard-to-obtain pathogen sequences is possible because the sequences are readily available in electronic public databases, including the sequence database GenBank (National Library of Medicine, 2022) from the National Institutes of Health (NIH). Community portals, such as National Center for Biotechnology Information (NCBI) Virus (National Library of Medicine, undated), provide a specialized interface for finding sequences of viruses, and Virus-Host DB (Mihara et al., 2016) provides information about relationships between viruses and hosts linked to NCBI taxonomies. The creation of viable pathogens or organisms using commercially available or synthesized gene products has been demonstrated in several publications. For example, Gibson et al. (2008) synthesized the entire genome of a bacterium, *Mycoplasma mycoides*, using commercially purchased synthetic sequences, and Gibson et al. (2010) inserted it into a related bacterium whose own genome had been removed, creating a new organism that could replicate and grow. Researchers have also synthesized viruses, including the poliovirus (Cello, Paul, and Wimmer, 2002), 1918 Spanish influenza virus (Tumpey et al., 2005), SARS-CoV-2 (Xie et al., 2021), and Mpox (Liu et al., 2023).

In 2016, Noyce, Lederman, and Evans synthesized viable infectious horsepox virus, an orthopoxvirus closely related to the virus responsible for smallpox, which was presumed to be extinct. Their resulting paper (2018) described using publicly available sequence data repositories to obtain the horsepox virus genetic sequence, and then using commercial DNA synthesis services to purchase, as Koblentz (2017, p. 623) described, "DNA fragments of approximately 30 Kbp [kilobase pair] each that collectively represented the entire 212 Kbp genome with some overlap between the fragments." Using established laboratory techniques (World Health Organization [WHO], 2016a), the researchers transfected viral DNA into cells already infected with another poxvirus containing viral proteins needed to transcribe the viral DNA and replicate the virus. These experiments, while conducted for scientific purposes, highlight the potential risks associated with the use of synthesized gene products to create highly virulent and contagious pathogens that could cause widespread harm or lead to a global pandemic.

Existing Regulatory Programs Have Limited Oversight of Nucleic Acid Sequences

Existing regulatory programs provide lists of dangerous agents and rules that limit the possession, use, and transfer of those agents, and they also control biological sequences that could be used to create some of those agents. The Federal Select Agent Program (FSAP) regulates potentially dangerous agents, including the genomes of RNA viral select agents for which the genomes of those agents can be directly translated into the protein constituents (FSAP, 2024). Hybrid or chimeric agents, even of select agent pathogens, are not necessarily classified as select agents unless they have specifically been designated as such (Wells and Parker, 2023a; Wells and Parker, 2023b). In the United States, the possession, use, or transfer of a select agent or toxin requires registration with FSAP and certification by the government, including a check of relevant records by the Federal Bureau of Investigation (FBI) (FSAP, 2022). In addition, the U.S. Commerce Control List regulates the transfer of some potentially dangerous organisms and genetic elements (Bureau of Industry and Security, undated). Title 18 of the U.S. Code includes federal criminal statutes involving biological agents and toxins, including criminal penalties for the possession of select agents by anyone who is not registered to possess them (U.S. Code, 2012). In addition, authorization from the HHS secretary is required to possess or synthesize variola (smallpox) virus, including "any derivative of the variola major virus that contains more than 85% of the gene sequence of the variola major virus or the variola minor virus" (U.S. Code, 2012). WHO has also issued recommendations limiting synthesis and use of variola DNA (WHO, 2016b). But researchers have highlighted serious problems with relying on taxonomic lists of dangerous agents (Millett et al., 2023, p. 522):

> Taxonomic lists are an excellent tool for controlling access to whole organisms, which is necessary for export controls and when possession of certain agents has been criminalized. . . . However, modern biotechnology increasingly makes use of parts of organisms to confer new traits to species, sometimes making the origin species of a single part a poor (or even misleading) proxy for risk assessment.

Pathogen Sequences May Be Designed to Not Resemble Existing Regulated Agents

Malicious actors might be able to evade regulations on select agents by disguising or modifying dangerous ordered nucleic acid sequences. Furthermore, it is increasingly possible to design novel pathogenic nucleic acid sequences that are sufficiently different from existing select agents so that they might not be detected by screening tools. Even in 2006, the National Science Advisory Board for Biosecurity (p. 8) wrote that "[i]t is now feasible to produce synthetic genomes that encode novel and taxonomically unclassified agents with properties equivalent to, or potentially more harmful than, current Select Agents." The National Academies of Sciences, Engineering, and Medicine (2018), in assessing risks for the U.S. Department of Defense's Chemical and Biological Defense Program, identified de novo synthesis of pathogenic viruses and bacteria as one of the most pressing biodefense challenges.

Declining Synthesis Costs May Contribute to Proliferation

Finally, market forces in the synthetic nucleic acid industry have contributed to both lower costs and the proliferation of commercial providers (Carlson, 2022). As costs per base pair drop below $0.10 and order volumes increase, fixed screening costs may occupy a larger share of providers' costs (Diggans and Leproust, 2019). Shrinking profit margins on gene-length and oligonucleotide synthesis could limit providers' willingness to invest in screening, although providers with well-established screening programs may bear the least additional burden under expanded screening guidelines. A report from the J. Craig Venter Institute highlighted the difficulty of guaranteeing screening of gene synthesis orders by providers:

> All of the factors mentioned above—declining dsDNA synthesis costs, greater international participation in the industry, potential decentralization of dsDNA production—will make it more difficult for DNA providers to adhere to the Guidance and more difficult for policy makers to maintain a consistent level of biosecurity screening. (Carter and Friedman, 2015, p. 16)

Related Risks of Benchtop Synthesis Hardware

Benchtop nucleic acid synthesis machines—sometimes called "DNA printers"—are also proliferating. These devices are sold for use in customers' own research lab facilities. Benchtop synthesis of short oligonucleotides (which requires labor-intensive assembly) has been possible for decades, but on-site, easy-to-use synthesis of gene-length dsDNA is now possible (Carter, Yassif, and Isaac, 2023). Benchtop synthesis presents additional biosecurity challenges because it may be possible to use these on-site devices to bypass screening procedures employed by mail-order synthesis providers. Of particular concern is unauthorized user modification ("jailbreaking") of synthesizers to enable unchecked synthesis of custom sequences, including those belonging to dangerous pathogens. Although jailbreaking may circumvent technical screening measures, customer screening approaches, such as those described in this report, might remain useful for screening purchasers of benchtop synthesis devices and customers of consumable inputs to such devices. Some of the findings and recommendations described in this report may apply to this growing market.

Securing Gene Synthesis a Priority for Government, Industry, and Researchers

Early Warnings and Governance Proposals

Before U.S. government guidance on securing gene synthesis emerged, several early authors—provider companies and biosecurity researchers—recognized the need for governance of nascent gene synthesis technologies and the growing synthetic biology industry. Aldhous (2005) issued one of the first warnings about the malicious use of made-to-order DNA services and Church (2005) offered a proposal for safe synthetic biology, including one of the first calls for the regulation of commercial DNA synthesis and screening of requested sequences against a database. A public draft of the Declaration of the Second International Meeting on Synthetic Biology warned that "despite several years of ad hoc encouragement and inquiries to different government agencies, not all DNA synthesis companies examine their orders for DNA sequences encoding hazardous biological systems" (International Meeting on Synthetic Biology, 2006, p. 2). The draft declaration called for industry adoption of guidelines for sequence screening. These early warnings and proposals did not distinguish between synthesis of oligonucleotides and gene-length sequences.

Around the same time, the National Science Advisory Board for Biosecurity (2006) issued a report that addressed biosecurity concerns related to the synthesis of select agents. The report recommended that HHS and the U.S. Department of Agriculture—the government agencies that manage the select agent program—develop guidance to synthesis providers regarding synthesizing select agents, screening orders that may represent sequences of select agents, and reporting suspicious orders.

Garfinkel et al. (2007) proposed options for the governance of synthetic genomics, including requiring providers to screen orders and sequences and retain records, while Bügl et al. (2007) presented a proposal for the oversight of commercial DNA synthesis by a group of academics, industry executives, and security experts. A report prepared by Gryphon Scientific for the U.S. Department of Homeland Security described the trade-offs of possible regulatory actions to reduce biosecurity risks associated with the synthetic nucleic acid industry (Casagrande et al., 2007). Nouri and Chyba (2009) issued a warning that increased access to gene synthesis technology and the expansion of the industry represented a security risk that required solutions inspired by nuclear nonproliferation. But two industry representatives, Minshull and Wagner (2009, p. 800), responded skeptically, writing that increased biosecurity screening of sequences might stifle innovation:

> If customers suspected that the confidentiality of their sequences might in any way be compromised, we would witness a withering of the enormous amount of innovation currently facilitated by synthesis; it could even compromise our ability to respond to potential pandemics.

Formation of Industry Consortia

Gene synthesis industry fears about burdensome biosecurity regulation may have spurred efforts to self-regulate, generating recommendations and policies for improving screening. Some of these efforts arose from biosafety and biosecurity concerns within the industry, and others were in response to government recommendations or requests for comment. In 2009, the International Association for Synthetic Biology (IASB) issued a code of conduct for best practices in gene synthesis to "secure the foundations of this fledgling field against abuse and to bring Synthetic Biology to its full potential" (IASB, 2009, p. 1). The code recommended that members take steps to screen sequences for similarity to known dangerous agents, screen customers for legitimacy, and report suspicious orders to relevant law enforcement authorities. The IASB code further specified that a human expert would investigate sequences flagged by automated sequence screening. Two synthesis providers, DNA2.0 (now ATUM) and GeneArt, disagreed and produced a code of conduct that did not require human review of flagged sequences (Check Hayden, 2009). In 2009, the resulting coalition of five providers became the International Gene Synthesis Consortium (IGSC). The IGSC published the "Harmonized Screening Protocol for Gene Sequence & Customer Screening to Promote Biosecurity," which specified sequence screening, customer screening, recordkeeping, order refusal and reporting, and regulatory compliance rules for member companies (IGSC, 2009). The IGSC eventually accepted the human review requirement for flagged sequences.

2010 U.S. Department of Health and Human Services Guidance and the Response from the Gene Synthesis Industry

In 2010, formal government guidance arrived: the Office of the Secretary at HHS issued "Screening Framework Guidance for Providers of Synthetic Double-Stranded DNA," representing the first comprehensive response to prior calls for formal guidance and providing clarity on the responsibilities of providers for screening gene synthesis orders (HHS, Office of the Secretary, 2010). The 2010 guidance laid out three components of screening: customer screening, sequence screening, and follow-up screening. Although the HHS guidance was nonbinding, it presented clear minimum standards for providers to follow and clarified reporting options when a provider identified a suspicious gene synthesis order. The HHS guidance recommends that providers check the legitimacy of customers and screen dsDNA orders over 200 base pairs in length but does not address the screening of synthetic oligonucleotide orders. The guidance includes recommendations for the review of sequences and detection of sequences of concern (SOCs), but does not specify whether this should be conducted by a human reviewer.

Industry calls to improve the security of synthesis services and compliance with federal guidance continued. In 2015, the J. Craig Venter Institute issued *DNA Synthesis and Biosecurity: Lessons Learned and Options for the Future* (Carter and Friedman, 2015), which described the 2010 HHS guidance, its implementation by providers at the time, the earlier IGSC rules for member companies, and recommendations for improving compliance with screening guidelines. The report recommends requiring U.S. federal grantees to obtain dsDNA only from companies that comply with HHS guidance. In 2017, the IGSC produced the second version of its Harmonized Screening Protocol, requiring IGSC member companies to screen sequences and customers, retain records, and report suspicious orders to law enforcement. The new IGSC rules were intended to "help ensure broad compliance with HHS Guidance for Double-Stranded DNA Providers and other international standards" (IGSC, 2017, p. 1) by strengthening the screening protocol employed by member companies. IGSC members are required to "screen the complete DNA sequence of every synthetic gene order," for which *synthetic gene* is defined as "[a] gene or other DNA fragment produced by Gene Synthesis, typically greater than 200 base pairs in length" (IGSC, 2017, pp. 1, 4). The new IGSC rules also require human review for flagged sequences:

> When automated screening identifies a potential pathogen or toxin sequence, the order is reviewed by a human expert using common IGSC screening criteria and is either accepted, accepted with a requirement for additional customer review, or rejected. (IGSC, 2017, p. 2)

The IGSC recommendations are significant because they expand on the 2010 HHS recommendations and make screening and compliance with these enhanced rules mandatory for IGSC member companies. However, IGSC has no mechanism to monitor or evaluate member companies' compliance with these rules.

DiEuliis, Carter, and Gronvall (2017, p. 2) presented options for strengthening screening protocols, highlighted the costs of screening for providers, and warned that the 2010 "HHS Screening guidance is becoming obsolete." Diggans and Leproust (2019) described recommendations for improving biosecurity within the industry and highlighted the role of the IGSC screening protocol in filling gaps in guidance provided by federal authorities.

Updated Government Guidance

Partially in response to continuing concern by industry representatives, researchers, and other stakeholders, in 2020, the Assistant Secretary for ASPR issued a notice of review and revision of the 2010 HHS screening framework guidance (HHS, Office of the Secretary, Assistant Secretary for Preparedness and Response, 2020). This request for comments to inform a revision of the existing guidance sought perspectives from industry and other life-sciences stakeholders on defining SOCs, sequence screening approaches, biosecurity measures, and customer screening.

Industry organizations contributed their perspectives. In 2022, the Engineering Biology Research Consortium (EBRC) issued considerations and recommendations for updating the 2010 HHS guidance, including proposals for improving technical sequence screening, strengthening customer screening, and implementing follow-up screening on flagged orders (EBRC, 2022). EBRC recommended specific customer screening best practices, including

> i) requiring customer identification data, including a physical, non-residential shipping address (PO boxes are not permitted given the lack of traceability); ii) restricted party screening to prevent any prohibited individuals or entities from receiving U.S. products or technology; iii) verifying a legitimate need—or, minimally, a scientifically sound use—and authorization before providing sequences for regulated pathogens or toxins; and iv) noticing "red flags" such as evasiveness around identity or affiliation. (EBRC, 2022, p. 6)

The American Society of Gene and Cell Therapy (Barrett, 2022) issued public comments on the proposed revisions and recommended expanding the scope of guidance to include single-strand DNA and dsDNA as short as 50 nucleotides in length, expanding the set of SOCs beyond the select agent and Commerce Control lists, and strengthening customer screening by requiring customers to preemptively verify their legitimacy when ordering synthetic SOCs. The American Society of Gene and Cell Therapy recommended against the screening of DNA and RNA oligonucleotides shorter than 20 nucleotides in length because of concerns that this would burden providers.

State Legislative Efforts

In the absence of federal legislation to secure commercial nucleic acid synthesis, some U.S. states have attempted to develop screening requirements and verification processes, often using the IGSC rules as a template. These efforts have been only partly successful. California Assembly Bill No. 70 proposed amending the state Health and Safety Code and requiring the California Department of Public Health to

> develop a verification process, with input from the IGSC and industry stakeholders, to verify that a gene synthesis provider and manufacturer of gene synthesis equipment adhere to customer and sequence screening protocols that are equivalent to, or stronger than, the IGSC's Harmonized Screening Protocol. (Assembly Bill No. 70, 2021)

The bill required any entity receiving state resources to purchase gene synthesis products only from providers that were current members of the IGSC or were otherwise verified. In addition, any gene synthesis providers operating in California that did not conduct IGSC screening or its equivalent would be fined. Although the bill was passed by the California State Assembly, it was vetoed by Governor Gavin Newsom, who wrote that "consideration should be given to whether a patchwork of state and federal regulations on biosecurity is the most effective way to approach an issue of international magnitude" (Newsom, 2021).

In 2022, a less stringent bill, California Assembly Bill No. 1963, was passed and enacted, requiring California State University and requesting the University of California to "develop systemwide guidance for purchasing gene synthesis equipment or gene synthesis products from gene synthesis providers who prevent the misuse of synthetic genes and safeguard the benefits of gene synthesis technology while minimizing risk, and to consider including IGSC criteria in their guidance" (Assembly Bill No. 1963, 2022).

In Maryland, House Bill 1256 would have required the Maryland Department of Health to "develop gene sequence and customer screening guidelines for gene synthesis providers and manufacturers of gene synthesis equipment" operating in or with Maryland and would have established civil penalties for failure to maintain certification (House Bill 1256, 2021). The bill would have required that "entities that receive State resources may purchase gene synthesis products and equipment only from certified providers and manufacturers," but the bill was withdrawn by its sponsor in 2021 (House Bill 1256, 2021).

Increasing Engagement and New Government Guidance

In 2022, the White House issued the *National Biodefense Strategy and Implementation Plan for Countering Biological Threats, Enhancing Pandemic Preparedness, and Achieving Global Health Security*, which increased the urgency for strengthening gene synthesis screening and called on the U.S. government to "[e]nhance screening of domestic biotechnology research, nucleic acid and biologics synthesis orders, and systems for identification and reporting suspicious orders, in consultation with relevant private sector partners" (p. ix).

Researchers and organizations also contributed responses to the HHS draft update. The Institute for Progress proposed improving capabilities for sequence screening, facilitating customer screening via licensing, automating the reporting of suspicious customers and orders to authorities, and withholding federal funding from entities that obtain synthetic genes from noncompliant providers (Williams and Kane, 2023). Hoffmann et al. (2023) offered proposals for improving technical sequence screening and SOC databases. Batalis et al. (2024) warned that new artificial intelligence (AI) tools in combination with made-to-order nucleic acid synthesis could enable malicious actors to hasten development of a biological weapon and recommended mandating screening for all providers. Carter, Yassif, and Isaac (2023) offered recommendations to secure benchtop synthesis devices, including enhancing customer screening to ensure the legitimacy of users, improving sequence screening, and establishing incentives for compliance.

In response to the public comments and growing agreement on the need to reform its earlier guidance, ASPR issued its "Screening Framework Guidance for Providers and Users of Synthetic Nucleic Acids" in 2023, a comprehensive revision to the 2010 HHS guidance (ASPR, 2023b). The new guidance recommended screening single-strand DNA and dsDNA sequences and implementing records management and cybersecurity measures. Among the changes to the previous guidance, the 2023 ASPR recommendations include specific match criteria for sequence screening, detailed criteria for customer legitimacy checks, and options for reporting to law enforcement (in particular, the local FBI weapons of mass destruction [WMD] coordinator).

International Initiatives

U.S. efforts to secure nucleic acid synthesis have not occurred in isolation. In 2022, the United Kingdom's (UK's) "Biosecurity Strategy: Summary of Public Response" called for regulatory action to address risks of increased access to gene synthesis: "Building international laws and restrictions to address concerns around the democratisation of science were also raised, where increasing access to DNA synthesis risked use by hostile actors and required regulatory oversight" (Cabinet Office, 2022). The Centre for Long-Term Resilience conducted a workshop for the UK government and provided recommendations for synthetic nucleic acid screening. In 2022, WHO issued the *Global Guidance Framework for the Responsible Use of the Life Sciences* and highlighted DNA synthesis in one of its risk scenarios.

In 2020, the World Economic Forum and the Nuclear Threat Initiative (NTI) proposed an international Common Mechanism for DNA Synthesis Screening and identified the need for a separate organization focused on the risks of emerging biotechnologies (World Economic Forum and NTI, 2020). In 2022, NTI Global Biological Policy and Programs (NTI | bio) hired Piers Millett to serve as the executive director of this biosecurity organization, the International Biosecurity and Biosafety Initiative for Science (IBBIS). In 2024, NTI | bio formally launched IBBIS as the long-term home for the Common Mechanism for DNA Synthesis Screening and for the Technical Consortium, a group of biosecurity experts from industry and government. The Common Mechanism is DNA synthesis screening software that is free, distributed, open-source, and automated, intended to increase the international adoption of synthesis screening and to make screening easier to implement for synthesis providers (Wheeler et al., 2024). IBBIS now operates independently, with two staff in addition to Millett and funding from philanthropic donors (IBBIS, undated; NTI, undated). Nongovernmental researchers and organizations have offered several proposals and products to help providers conduct sequence screening, including SeqScreen (Balaji et al., 2022), Aclid (undated), FAST-NA (Raytheon BBN, undated), UltraSEQ (Battelle, undated), and SecureDNA (undated).

Executive Order 14110

In October 2023, President Joseph R. Biden issued Executive Order 14110, which tasks the Director of the White House Office of Science and Technology Policy (OSTP) with establishing a framework "to encourage providers of synthetic nucleic acid sequences to implement comprehensive, scalable, and verifiable synthetic nucleic acid procurement screening mechanisms, including standards and recommended incentives" (p. 75201). The executive order includes a mechanism that incentivizes compliance with the new screening rules by requiring that "all agencies that fund life-sciences research shall, as appropriate and consistent with applicable law, establish that, as a requirement of funding, synthetic nucleic acid procurement is conducted through providers or manufacturers that adhere to the framework" (Executive Order 14110, 2023, p. 75202).

The executive order also requires OSTP to develop methods for ensuring that providers comply with the new synthesis guidelines:

> determine standardized methodologies and tools for conducting and verifying the performance of sequence synthesis procurement screening, including customer screening approaches to support due diligence with respect to managing security risks posed by purchasers of biological sequences identified in subsection 4.4(b)(i)(A) of this section, and processes for the reporting of concerning activity to enforcement entities. (Executive Order 14110, 2023, p. 75201)

The *Framework for Nucleic Acid Synthesis Screening* (Fast Track Action Committee on Synthetic Nucleic Acid Procurement Screening, 2024, p. 5) incorporates and supplements the 2023 ASPR guidance and applies to "federally funded life sciences research involving the procurement of all types of synthetic nucleic acids—including but not limited to DNA and RNA, whether single- or double-stranded, as well as whole organism genomes (e.g., viruses, bacteria)—containing any synthetic SOC." The framework outlines the following six responsibilities of providers:

- attest to implementing the screening framework
- screen purchase order for synthetic nucleic acids to identify SOCs
- screen customers submitting purchase orders of SOCs
- report potentially illegitimate purchase orders involving SOCs
- retain records
- ensure cybersecurity and information security.

The framework also specifies that sequences at least 200 nucleotides long should be screened for SOCs, with this threshold reduced to 50 nucleotides in 2026:

> At a minimum, DNA or RNA, single- or double-stranded, 200 nucleotides (including the corresponding amino acid sequence, if applicable) or longer should be screened for SOCs. As of October 13, 2026, this screening window will be decreased to 50 nucleotides, and Providers should implement screening mechanisms that detect the potential for shorter nucleotides to be assembled into SOCs when multiple synthetic nucleic acids are ordered by the same customer in a bulk order or for multiple orders over time. (Fast Track Action Committee on Synthetic Nucleic Acid Procurement Screening, 2024, pp. 6–7)

Executive Order 14110 and the framework developed by the Fast Track Action Committee on Synthetic Nucleic Acid Procurement Screening make screening required for providers who receive orders from customers using federal funding. In implementing the specific requirements of the executive order, government agencies will develop and standardize a framework for screening that could be adopted by all or most providers or required by industry consortia, such as IGSC. More broadly, the screening framework developed under the executive order could inform such international efforts as the Common Mechanism for DNA Synthesis Screening. Although the executive order specifically restricts U.S. federal government funding to providers that comply with the framework, many elements of the framework are not specific to the United States. All the recommendations in this report could be modified for the international context, including the overall screening workflow, the establishment of a third-party database to log information about orders, the implementation of customer screening methods, communication with relevant law enforcement entities, and assessment of conformity.

In this report, we focus our recommendations on several components of the executive order that are related to the screening of synthetic nucleic acid orders: an overview of the screening framework to be implemented by providers, the establishment of orders, the implementation of customer screening, communication with law enforcement, and assessment of conformity. We focus on these components for two primary reasons. First, we are aware of a variety of industry efforts to provide technical sequence screening as a service, including SeqScreen (Balaji et al., 2022), Aclid (undated), FAST-NA (Raytheon BBN, undated), UltraSEQ (Battelle, undated), and SecureDNA (undated). For this reason, we do not provide detailed recommendations on technical sequence screening, nor do we recommend a minimum sequence length above which all ordered synthetic sequences must be screened. Second, our review of the literature summarized in this chapter indicates that there is a significant need for the development of an overall screening framework with detailed criteria, a mechanism for detecting split orders across providers, better tools for customer screening, clearer criteria for reporting to law enforcement, and a mechanism for ensuring providers' conformity with the framework.

CHAPTER 2

Research Approach

Scope and Intended Audience

The purpose of this report is to provide recommendations that describe ways the U.S. government can implement components of Executive Order 14110 (2023) related to synthetic nucleic acid sequence screening, customer screening, and conformity assessment. The specific components of the executive order addressed in the findings and recommendations of this report (Chapters 3 and 4, respectively) are as follows:

- 4.4(b)(i)(B) "determine standardized methodologies and tools for conducting and verifying the performance of sequence synthesis procurement screening, including customer screening approaches to support due diligence with respect to managing security risks posed by purchasers of biological sequences . . . and processes for the reporting of concerning activity to enforcement entities" (p. 75201)
- 4.4(b)(ii)(A and D) "develop and refine for possible use by synthetic nucleic acid sequence providers: (A) specifications for effective nucleic acid synthesis procurement screening . . . and (D) conformity-assessment best practices and mechanisms" (p. 75202)
- 4.4(b)(iv)(A) "develop a framework to conduct structured evaluation and stress testing of nucleic acid synthesis procurement screening" (p. 75202).

We focus our recommendations on several specific components of a proposed screening protocol, including customer screening, communication with law enforcement about potentially dangerous or illegitimate orders, and assessment of provider conformity with the proposed framework. In providing recommendations for these components, we also describe the broad outlines of the screening workflow and options for keeping records about customer orders. We do not provide detailed recommendations regarding the technical aspects of sequence screening, the curation of a database containing SOCs, or the process of securing benchtop gene synthesis hardware. Although we note cases in which historical recommendations have made a distinction between screening requirements for short oligonucleotide sequences and longer gene-length sequences, we do not propose a length threshold below which screening need not be conducted.

The intended audience for this report is officials designated by the executive order to lead the establishment of the nucleic acid synthesis screening framework, specifically "the

Director of OSTP, in consultation with the Secretary of State, the Secretary of Defense, the Attorney General, the Secretary of Commerce, the Secretary of Health and Human Services (HHS), the Secretary of Energy, the Secretary of Homeland Security, the Director of National Intelligence, and the heads of other relevant agencies as the Director of OSTP may deem appropriate" (Executive Order 14110, 2023, p. 75201). In addition, the recommendations presented in this report may be useful to officials tasked with engaging with stakeholders: "the Secretary of Commerce, acting through the Director of NIST [National Institute of Standards and Technology], in coordination with the Director of OSTP, and in consultation with the Secretary of State, the Secretary of HHS, and the heads of other relevant agencies as the Secretary of Commerce may deem appropriate," as well as the Assistant to the President for National Security Affairs, who is tasked with coordinating the process of reviewing federal funding requirements (Executive Order 14110, 2023, pp. 75201–75202). Finally, the report may be useful to officials tasked with reporting the results of the conformity assessment: "the Assistant to the President for National Security Affairs, the Director of the Office of Pandemic Preparedness and Response Policy, and the Director of OSTP" (Executive Order 14110, 2023, p. 75202).

Literature Review

We conducted an exploratory literature review between January and March 2024 to understand the current state of commercial gene synthesis screening, government and industry screening guidelines, and proposals for improving screening. We reviewed historical and current U.S. government guidance documents and conducted Google and Google Scholar searches for research articles on synthetic biology, DNA synthesis screening, and engineered pathogens. We reviewed scholarly research on securing synthetic biology and nucleic acid synthesis screening, as well as gray literature reports, commentaries, and recommendations by industry researchers. We also reviewed U.S. state legislative proposals and international mechanisms for synthetic nucleic acid screening. We sought to understand the common features of existing proposals and guidelines for securing nucleic acid synthesis screening and identify gaps in current screening protocols. Literature sources included the following:

- provider statements and advertised services, including prices
- U.S. government guidance documents and requests for comment
- U.S. state legislation
- industry consortium rules
- industry recommendations, proposals, and comments
- findings and recommendations from academic, industry, and other nongovernmental researchers
- international organization recommendations.

Research Questions

To address the components of Executive Order 14110 described above, we formulated the following research questions:

- What is the current state of commercial gene synthesis screening?
- How should providers determine whether an order is suspicious?
- How can providers or third parties ensure that suspicious orders split across providers will be detected?
- How should information about suspicious orders be relayed to law enforcement?
- How should providers' conformity with screening rules be verified?

CHAPTER 3

Findings

In this chapter, we present the findings from our review of the relevant literature. We emphasize findings that are most relevant to our research questions: assessment of biosecurity risk arising from current screening guidance and practices and opportunities for improving screening under Executive Order 14110. These findings build on the background material presented in Chapter 1, especially the history of government and industry guidance on screening, historical examples of biosecurity risks arising from synthetic nucleic acid sequences, and calls for an improved screening framework. The findings also lay the groundwork for recommendations on implementing the requirements of the components of Executive Order 14110 described in Chapter 2. We briefly summarize the findings as follows:

- Commercial gene synthesis presents a serious risk of misuse or accident.
- Screening protocols vary across providers.
- Providers currently have no way to share information about suspicious customers or orders.
- Strengthening customer screening and verification is a promising way to help secure gene synthesis.
- Clearer standards for reporting suspicious orders to authorities would streamline follow-up screening and improve enforcement.
- The U.S. government currently has no mechanism for assessing the conformity of providers' screening systems with guidelines or policies.

Commercial Gene Synthesis Presents a Serious Risk of Misuse or Accident

The synthetic nucleic acid industry has contributed to innovation in the biological sciences by reducing costs and increasing access to custom-made synthetic DNA/RNA constructs. However, affordable, accurate, and accessible nucleic acid synthesis could permit individuals or groups with limited financial resources, fewer personnel, less advanced laboratory equipment, and lower levels of education or expertise to obtain genetic elements of a dangerous pathogen. In addition to intentional misuse, synthetic genes can be used to construct potentially dangerous agents for use in legitimate biological research, such as the development of a new vaccine. If potentially dangerous pathogen sequences are provided to unauthorized individuals or laboratories that do not meet safety standards, the intentional or unintentional release of a dangerous pathogen could occur.

These risks are not speculative—the resurrection of a virus from a synthetic nucleic acid sequence is now possible. Several examples are cited in Chapter 1, including *Mycoplasma mycoides* (Gibson et al., 2008; Gibson et al., 2010), poliovirus (Cello, Paul, and Wimmer, 2002), the 1918 Spanish influenza virus (Tumpey et al., 2005), SARS-CoV-2 (Xie et al., 2021), Mpox (Liu et al., 2023), and horsepox (Noyce, Lederman, and Evans, 2018). Koblentz (2017, p. 623) writes of the synthesis of horsepox that

> [t]he entire project was conducted using only "commercially available information, technology and tools" and took Evans's lab 6 months to complete at a cost of about $100,000. The most time-consuming part of the project was the amount of time it took the DNA synthesis company Geneart to construct the DNA fragments.

A WHO report on the synthesis of horsepox found that "recreation of such viral genomes did not require exceptional biochemical knowledge or skills, significant funds or significant time" (2016a, pp. 29–30).

Although cost, equipment, and expertise are still barriers to de novo synthesis of viable pathogens, commercial synthetic nucleic acid synthesis could hasten the development of a biological weapon or a highly pathogenic organism. Writing about what the synthesis of horsepox by Noyce, Lederman, and Evans (2018) means, Koblentz warned that "a disgruntled or radicalized scientist, sophisticated terrorist group, unscrupulous company, or rogue state" could create and release a pathogen with the potential to cause mass death (2017, p. 621).

Screening Protocols Vary Across Providers

Synthetic nucleic acid providers do not want their products to be used to create a biological weapon or other dangerous agent (Diggans and Leproust, 2019; Kane and Parker, 2024), but currently, there is no unified screening process (National Security Commission on Emerging Biotechnology, 2024b). Ensuring that all providers effectively screen customers and sequence orders has been challenging, and the lack of clear standards for the screening of synthetic nucleic acid orders has resulted in a patchwork of screening solutions employed by providers. Most U.S. providers are members of IGSC, which requires member companies to follow its Harmonized Screening Protocol, but IGSC has no mechanism for verifying compliance. Implementation of screening in compliance with the new 2023 ASPR guidelines remains optional and still leaves substantial room for variation in screening practices (Kane and Parker, 2024). Variation in screening practices is not necessarily a biosecurity risk—providers may accomplish the same overall goals through different screening methodologies. But some providers' screening protocols may be easier for a nefarious actor to bypass. Kane and Parker (2024, p. 2) write that

> the lack of regulation for sequence and customer screening has naturally led to inconsistencies throughout the field, some of which might be exploitable by bad actors. Indeed, many stakeholders have expressed the opinion that there is a need for a universal screening mechanism and improved governance of the synthetic DNA field.

EBRC (2022, p. 2) describes the link between unclear government guidance and potential harms as follows:

> [T]he IGSC suggests that its members represent approximately 80% of global DNA synthesis capacity. It therefore is likely that the majority of synthetic DNA produced globally is synthesized by providers that adhere to the Guidance and/or the IGSC Protocol, but a nontrivial amount of synthetic DNA is likely produced and shipped from companies that have no, or inadequate, screening mechanisms in place. Those companies that do not screen are at risk of inadvertently violating national regulations and/or international agreements by unknowingly shipping prohibited sequences domestically or internationally. They might also indirectly contribute to nefarious activities through the synthesis of non-regulated sequences with clear potential for misuse for questionable customers. For those that do screen, outdated Guidance leaves providers of synthetic DNA with a lack of clarity on key elements of screening. . . . As such, individuals without the training, need, and/or authority to handle sequences that could be used to cause harm may be able to access and use them—intentionally or not—to damage human and/or environmental health and well-being.

Providers Currently Have No Way to Share Information About Suspicious Customers or Orders

On the basis of our review of government guidance, we assess that there is currently no way for providers to detect orders for dangerous sequences that are split across multiple providers or orders that are split across multiple customer names. A malicious actor seeking synthesis of a dangerous agent might split the desired sequence into orders that are sent to several providers, each of which may not appear concerning in isolation. Providers also have no formal and universal way of sharing information about rejected orders or suspicious customers. Proposals for improving information-sharing involve the creation of an independent body to conduct screening across orders to different providers or to facilitate sharing information about suspicious orders with law enforcement agencies. EBRC (2022, p. 8) recommends information-sharing between providers to identify customers or orders that have been flagged or denied in the past and describes the creation of an independent entity to conduct customer and sequence screening:

> [U]pdated Guidance should consider if or how [the U.S. government] should encourage information sharing between companies. . . . An alternative model that would take significant investment to establish but that might ultimately improve the safety and

> security of the synthetic DNA synthesis market is the establishment of an independent third party used by all providers of synthetic DNA operating in the United States for customer and sequence screening.

An Institute for Progress report recommended direct information-sharing with authorities, even in the absence of suspicious order characteristics (Williams and Kane, 2023).

Existing proposals would create new federal entities, charge existing entities with keeping such a database of order records, or require providers to share all or most information directly with law enforcement. But these approaches may not be feasible and may not be acceptable to customers who consider the sequences they order to be highly proprietary and would be concerned about sharing that information beyond the synthesis provider. Furthermore, these proposals could impose undue burden on providers and may subject customers to unnecessary government scrutiny.

A third-party entity—either a private company or a consortium of industry stakeholders—could maintain a log of order information collected by providers during an improved customer screening process, against which new orders could be checked. Prior flagged orders by a given customer present in such a database might encourage providers to scrutinize future orders from the same customer more carefully. If a provider reports a suspicious customer order to law enforcement, the log would contain records of prior orders from the same customer, which could be obtained by authorities for investigative purposes. A system for checking new orders against historical orders by the same customer to other providers might be able to identify such a pattern.

The ultimate goal of law enforcement may not be to deny access to a specific resource but rather to prevent the completion and deployment of a malicious project. In this case, the largest source of value from reporting and analyzing suspicious customer activity may not be informing order fulfillment but providing an indication that a malicious project is underway and that investigation by law enforcement is warranted. Additionally, although a single suspicious order might be insufficient to justify such action, patterns may be detectable across orders, which when taken together, could clear a threshold for investigation.

Strengthening Customer Screening and Verification Is a Promising Way to Help Secure Gene Synthesis

Customer screening is an underappreciated way of ensuring that malicious actors do not obtain dangerous biological materials (Alexanian and Carter, 2024; Carter, 2024). In a review of screening practices, Kane and Parker (2024) found highly variable customer screening practices across providers. But enhanced customer screening requirements could make gene synthesis services more secure and might provide an additional layer of protection that technical sequence screening cannot. Carter (2024, p. 4) writes that the goal of customer screening is to

> reduce the risk that someone deliberately or irresponsibly misuses life sciences products, services, or infrastructure to cause harm. An effective customer screening framework will make it harder for individuals or small groups without legitimate credentials to misuse these capabilities. It also reduces the risk that a researcher accesses potentially harmful reagents or tools that they have no reason to use.

The framework developed by the Fast Track Action Committee on Synthetic Nucleic Acid Procurement Screening (2024, p. 4) lists "[s]creen customers submitting purchase orders of synthetic nucleic acids with SOCs" as one of six provider responsibilities. However, SOC-based screening approaches have several weaknesses (Wheeler et al., 2024) and may fail to detect sequences that could be used to construct a dangerous pathogen. Furthermore, short oligonucleotide sequences can be difficult to screen but might be assembled by a customer into an SOC (HHS, Office of the Secretary, 2010). By limiting customer screening to orders that have been identified as containing SOCs, the value of low-cost automated customer checks to detect competent nonstate groups and individuals may be substantially reduced. In this way, the customer screening requirement specified by the framework may be too narrow to effectively mitigate risks from novel biological threats and sophisticated nonstate actors, including technically competent individuals. Additionally, if providers are required to screen only customer orders that they know contain an SOC, this may increase costs for providers whose systems are more sensitive in identifying potential SOCs.

Tools and Standards for Customer Screening

A variety of tools for automated customer screening are commercially available and in use in other industries, such as finance (Carter, 2024). However, many providers have not yet adopted automated tools to verify customer identity or to screen customers against sanctions lists. Current IGSC requirements call for gene synthesis customers to identify themselves to providers when ordering, but many providers take only limited steps to verify the identity of the individuals (Kane and Parker, 2024). Some providers conduct searches of the scientific literature, contact customers, or manually check addresses against public or commercial databases. However, these approaches involve subjective judgments, demand staff time, and can delay order processing, creating a compliance burden for providers.

There is substantial existing government and private sector infrastructure for verifying customer identities. NIST has developed Digital Identity Guidelines specifying the requirements for different levels of identity assurance (Grassi, Garcia, and Fenton, 2020) and a risk assessment process for determining the appropriate levels of identity assurance (NIST, 2024). Risk analysis processes could be conducted for distinct identity assurance use cases. Although the NIST guidelines have been developed for use by government agencies, the standard states that "private sector organizations and state, local, and tribal governments whose digital processes require varying levels of assurance may consider the use of these standards where appropriate" (Grassi, Garcia, and Fenton, 2020, p. 4). Given the potential impacts to public health and national security that are associated with malicious and accidental pathogen development and release that are potentially enabled by nucleic acid products, this risk assessment framework may be more appropriate than those focused on private sector impacts. NIST allows the certification of commercial entities to specific identity assurance levels by approved third-party assessors, and the availability of automated systems adopted by providers may be considered when determining requirements (Kantara Initiative, undated). Automated and robust systems for domestic identity verification that comply with NIST standards are commercially available (Jumio, undated; Proof, undated; Socure, undated).

Use of the NIST guidelines for identity verification would permit providers to impose different identity assurance levels for various users of synthetic nucleic acid sequences. For example, verification of the identity of a principal user (the person who places a synthetic sequence order and oversees its use in the laboratory) may require more-stringent identity assurance standards than an end user (the individual who ultimately receives and uses the synthetic sequences). Similarly, accounts that routinely order potentially dangerous sequences on behalf of biosafety level 4 facilities may justify more-stringent identity verification standards than other users. The outputs of this process may enable better-informed balancing of risk reduction against provider compliance burden.

Identity Verification

Government-issued identification documents are a promising means of identity verification, including for international customers. Commercial systems that recognize documentation from many countries are available, which reduces the impact of provider identification verification requirements on the accessibility of services for international customers (ID Analyzer, undated; Keesing Technologies, undated). The submission of fake or stolen government-issued identity documents can be detected by these systems, which may result in serious consequences for malicious actors, providing incentives for customer compliance and increasing enforceability. Although generative AI deepfakes can fool optical identification verification systems (iProov, 2024), deepfake generation tools can become ineffective as new detection techniques are developed (Agarwal et al., 2020; Yu et al., 2024). Stolen identity documents may be used to place orders using a false name, but customer screening software used in finance and other industries that are subject to know-your-customer regulations include fraud detection features designed to detect documents that are publicly available on the internet or otherwise have risk indicators for fraud (ID Analyzer, undated). These document fraud detection efforts are supported by government initiatives, such as the Swiss Interstate Database of Fraudulent Identity Documents (Moulin, Weyermann, and Baechler, 2022). When a stolen document is detected and screened by commercial identity verification solutions, it can be flagged as fraudulent. Storage and periodic rescreening of previously submitted identity verification data can allow for new tools to detect historical fakes. This practice may reveal to a provider that a previously approved order was placed using fraudulent information, in which case, other data associated with the order, including company name, delivery address, phone number, internet protocol (IP) address, and sequence information, may be flagged as suspicious.

The IGSC Harmonized Screening Protocol already recommends that providers collect much of this information. Checking customer and user names against databases of sanctioned or otherwise prohibited organizations and individuals is also a powerful way of preventing known bad actors from ordering synthetic nucleic acids. Commercial services exist to conduct checks against screening lists, enforcements, and court filings around the world (LexisNexis Risk Solutions, undated). Some customer practices could limit the effectiveness of identity verification. For example, verified principal or end users could share nucleic acid products with unverified users. However, ensuring that there is at least one verified identity associated with an order could still provide a starting point for investigation in cases for which this is warranted.

Checking for Identity Obfuscation Methods

During customer screening, automated checks for identity obfuscation methods may yield significant security benefit at low compliance cost. The IGSC Harmonized Screening Protocol requires member companies to refuse orders to post office boxes, mail forwarding services, and residential addresses (IGSC, 2017). Tools for automated address verification are commercially available (Experian, undated), although these tools may give less reliable results for international customers. Tools for checking whether web traffic is originating from The Onion Router (Tor), commonly known as the dark web, are also widely available (Falco, undated; WatchGuard, undated). Requiring providers to check and report orders originating from Tor exit nodes could help secure synthesis against a system commonly used to evade law enforcement oversight of online activity. Although we are not aware of any gene synthesis providers that accept difficult-to-trace cryptocurrency, such as Monero (undated), as payment, requiring providers to refuse anonymous forms of payment that are commonly associated with law enforcement evasion may be a promising and low-cost approach to securing gene synthesis against misuse by customers who seek to obscure their identity.

Usefulness of Customer Screening When Technical Sequence Screening May Fail

Technical sequence screening has the potential to identify sequences that are similar to those of known pathogenic organisms. However, these measures may not be sufficient to guard against novel organisms whose genomic sequence differs from those of existing pathogens. Sequences produced using generative AI foundation models, such as those capable of protein or whole genome modeling (Ingraham et al., 2023; Nguyen et al., 2024), may be particularly difficult to identify based on homology to known SOCs. Likewise, screening for dangerous sequences ordered from mail-order providers may be undermined by the proliferation of benchtop synthesizers because the devices under the physical control of the user might be compromised. Malicious (and benign) actors may reverse engineer proprietary hardware in their possession (Lewis, 2015; McLoughlin, 2008; Zaidenberg, 2020). Because benchtop synthesizers might not receive security updates once they are delivered to the customer, securing benchtop synthesis would require that all devices are secured against future hacking attempts at the point of sale. Although some manufacturers have developed business models in which customers are locked into their specific consumables (Kane and Parker, 2024), it is unclear whether manufacturers will be able to secure the hardware against hackers who seek to circumvent such controls to synthesize dangerous sequences.

The vulnerability of sequence screening—particularly when initiated or conducted by software running on benchtop synthesis devices—has strategic implications for the relative prioritization of developing technical and customer screening processes. If technical screening can be circumvented either by novel sequence generation or by the unrestricted use of synthesis devices, the role of point-of-sale customer screening grows in relative importance and can be extended to detect signs of misuse across a variety of biotechnology use cases. These cases may include purchases of equipment, organisms, or laboratory consumables. The repeated procurement of proprietary reagents and other distinctive materials with suitable customer screening may generate information that increases opportunities to disrupt malicious projects.

Clearer Standards for Reporting Suspicious Orders to Authorities Would Streamline Follow-Up Screening and Improve Enforcement

A lack of clarity about when and how providers must report suspicious orders to law enforcement agencies has resulted in widely varying policies and procedures. There is no uniform and codified mechanism for alerting law enforcement officials about suspicious synthetic nucleic acid sequence orders (National Security Commission on Emerging Biotechnology, 2024b). In a study of providers' screening practices, Kane and Parker (2024, pp. 8–9) warn that

> [t]he greatest vulnerability identified by this study is the surprisingly small number of providers with formalized procedures for law enforcement reporting in the case of a potentially dangerous order. It is true that many respondents have never had to report an order to law enforcement before, but this obviously does not preclude such an event from occurring in the future . . . Thus, it is critical that U.S.-based providers at the very least establish a relationship with their local FBI Weapons of Mass Destruction Coordinator. . . . In the long-term, governments must establish clear lines of communication with synthetic DNA providers such that there is no ambiguity in which cases must be reported to law enforcement.

Providers' varied reporting practices may result from the lack of definitive guidance about when providers must report information about a suspicious customer or order to authorities.

Historically, guidance to providers came in the form of nonbinding recommendations with vague or unspecified criteria for reporting. The 2010 HHS guidance did not require reporting suspicious orders and instead encouraged seeking advice from relevant entities:

> If the provider encounters a scenario where they would benefit from additional assistance in assessing an order, the provider is encouraged to seek advice from the relevant U.S. Government Departments and Agencies by contacting the nearest FBI Field Office Weapons of Mass Destruction (WMD) Coordinator. (HHS, Office of the Secretary, 2010, p. 62825)

When follow-up screening failed to resolve issues, the 2010 HHS guidelines also recommended that providers contact the Federal Select Agent Program, which is managed jointly by the Centers for Disease Control and Prevention and the Animal and Plant Health Inspection Service, or the Department of Commerce for international orders of concern.

In 2022, EBRC (2022, p. 8) described options for follow-up screening involving consultation with the FBI and also highlighted the cost penalties of mistakenly flagging legitimate orders, which would inappropriately trigger follow-up screening:

> [F]ollow-up screening can involve attempts to verify a principal user's identity, making contact with relevant institutional biological safety officers or research directors, a literature review for related or relevant publications, and, if necessary, consultation with the Weapons of Mass Destruction (WMD) Coordinator at the nearest FBI Field Office. This highlights the importance of minimizing false positives in screening and of supporting tool development that enables small companies to implement screening systems without incurring an overly arduous financial burden.

Soon after, the ASPR screening framework guidance (ASPR, 2023b, pp. 2–3) used stronger language and more specific criteria for reporting:

> If follow-up screening does not resolve concerns about the order, or if there is reason to believe a customer may intentionally or inadvertently violate U.S. laws or regulations, Providers should not fulfill the order and should contact designated entities within the U.S. government (i.e., U.S. Department of Commerce, Federal Bureau of Investigation [FBI]) for further information and assistance.

Still, the ASPR guidelines do not compel providers to screen orders, and the criteria for reporting to law enforcement are left unspecified beyond "unresolved concerns." The ASPR guidance is accompanied by a companion guide with a section titled "Scenarios Addressing Red Flags" for providers. These scenarios describe circumstances in which "the provider or manufacturer should contact the nearest FBI Field Office's WMD Coordinator and/or follow up with DOC [Department of Commerce], as appropriate" (ASPR, 2023a, pp. 9–10). But the companion guide does not provide explicit criteria for reporting in other settings, and some scenarios may not appear suspicious to providers. For example, the lack of clarity about the customer or the nature of the facility to which an order is to be shipped may warrant rejection of the order but may not be serious enough to *require* the provider to report the customer to the FBI.

In contrast to the voluntary guidance from federal authorities, the IGSC Harmonized Screening Protocol requires screening by member companies and provides general recommendations for reporting. IGSC requires member companies to have "established relationships with local and national law enforcement and intelligence authorities" and requires members to refer to law enforcement "any request for a gene associated with the pathogenicity of an organism received from a suspicious potential customer and/or potential customer failing to establish their legitimacy" if follow-up screening does not resolve concerns (IGSC, 2017, p. 3). The IGSC Harmonized Screening Protocol requires reporting to law enforcement, but it does not offer specific criteria that providers must use when making the decision to report. The IGSC provides a suspicious order reporting form, but Alexanian and Nevo (2024, p. 13) report that

> [t]he IGSC's suspicious order report infrastructure, which was designed to prevent customers from venue shopping for refused orders, has only been used once, and an industry contact speculated that may be because the reporting form was too detailed and time-consuming, and users were unaware they did not need to supply information in every field.

Alexanian and Nevo (2024) identify support for follow-up screening and law enforcement reporting as a major gap in current synthesis screening guidelines and practices. They propose supporting the creation of decision-support tools for follow-up screening of flagged orders, standardization of data formats for records of flagged orders, and required reporting of suspicious orders to law enforcement.

Some providers may be hesitant or unwilling to report suspicious orders to law enforcement because they seek to protect customers' proprietary sequence information. Carter and Friedman (2015, p. 8) describe the IGSC reporting rules while highlighting providers' desire to maintain confidentiality for their customers:

> The Guidance frequently suggests that providers contact the FBI or the U.S. Department of Commerce for additional advice. However, because confidentiality often is critical to purchasers of DNA for reasons of commercial competitiveness, providers of synthetic DNA take customers' confidentiality concerns very seriously.

The absence of stronger language mandating or recommending reporting may reflect providers' concerns about subjecting themselves or their customers to law enforcement scrutiny or reveling customers' intellectual property.

Executive Order 14110 calls for the U.S. government to address the lack of clarity in current gene synthesis screening guidelines specifically related to reporting to law enforcement. Section 4.4(b)(i)(B) tasks OSTP, in consultation with other federal agencies, with developing "processes for the reporting of concerning activity to enforcement entities" (Executive Order 14110, 2023, p. 75201).

The U.S. Government Currently Has No Mechanism for Assessing the Conformity of Providers' Screening Systems with Guidelines or Policies

The synthetic nucleic acid industry is currently self-regulated by providers and stakeholders. The 2023 ASPR guidance strengthened the earlier 2010 HHS guidance but still remains nonbinding. Executive Order 14110 compels relevant U.S. government agencies to create a framework for sequence and customer screening and to prohibit use of federal funding to purchase synthetic nucleic acids from providers that do not comply with guidelines. But the government currently has no way to verify that providers are in conformity with the new guidelines. Executive Order 14110 (2023, p. 75202) requires that "conformity-assessment best practices and mechanisms" be defined to "develop a framework to conduct structured evaluation and stress testing of nucleic acid synthesis procurement screening." The *Framework for Nucleic Acid Synthesis Screening* (Fast Track Action Committee on Synthetic Nucleic Acid Procurement Screening, 2024) requires only that providers self-attest that they are in conformity with the framework and makes no provisions for auditing or other kinds of compliance testing. The monitoring and evaluation of screening procedures take different forms among providers, when these procedures exist at all (Koblentz, 2017; National Academies of Sciences, Engineering, and Medicine, 2018). Although some providers evaluate their procedures on a random or regular basis, others may not.

Red teaming, attempting to create orders that would purposely violate synthesis screening guidelines, is an effective means of assessing the evolving risk of vulnerabilities in screening systems (Diggans and Leproust, 2019; Edison and Esvelt, 2024; Zhang and Gronvall, 2020). The U.S. government already conducts red team exercises to enhance critical infrastructure security. For example, the Cybersecurity and Infrastructure Security Agency (2023) performs red team assessments to help U.S. companies improve their resistance to cyberattack. Although there are synthetic DNA providers and stakeholders who conduct randomized or regular red teaming and run test sets as a form of evaluation protocols, many providers report that their standard review of flagged sequences during the normal ordering process is sufficient to monitor the effectiveness of their screening procedures (Kane and Parker, 2024). However, any false negatives a system produces might be missed by this sort of review. For this reason, regular red teaming and test set challenges represent a more comprehensive strategy for monitoring and evaluation.

Red teaming may also have a secondary value: Sequences shown to bypass a screening system can then serve as effective test cases during subsequent screening development once vulnerabilities have been patched. Regular testing of the screening system against an updated set of test sequences could ensure that the system still successfully detects all historical instances of flagged orders. Koblentz (2017) suggests that the IGSC should require its members to conduct annual exercises to test their customer and sequence screening procedures and commission third-party audits of their compliance with the consortium's code of conduct.

CHAPTER 4

Recommendations

This chapter describes recommendations for securing nucleic acid synthesis under Executive Order 14110. Each recommendation corresponds with at least one finding described in Chapter 3 and addresses a specific component of the executive order. The first recommendation proposes that an overall screening workflow be implemented by providers of synthetic nucleic acid sequences. This overall workflow contains several screening components described by subsequent recommendations, including customer screening, checking a database for evidence of split orders, and communicating with law enforcement about suspicious orders. The final recommendation covers assessment of providers' conformity with the proposed screening framework. We briefly summarize the recommendations as follows:

- Revise the 2023 ASPR guidelines to include specific information to be collected and decision criteria for each step of the screening pipeline.
- Establish a third-party database to log information about synthetic nucleic acid orders and periodically check records for evidence of suspicious activity.
- Establish customer screening requirements, including identity verification measures.
- Establish criteria and data requirements for sharing information about suspicious orders with law enforcement agencies.
- Implement procedures and standards for assessing the conformity of providers' screening practices with requirements.

Revise the 2023 ASPR Guidelines to Include Specific Information to Be Collected and Decision Criteria for Each Step of the Screening Pipeline

We recommend that the 2023 ASPR guidance be revised to describe the required screening steps for receipt of federal funding under Executive Order 14110. In our review of existing recommendations, screening protocols, and proposals, we found that protocols vary among providers in part because of varying standards and government guidance that does not specify in detail how providers should screen orders. We therefore recommend an overall screening workflow based on IGSC rules and the 2023 ASPR guidance that specifies the data collection and decision responsibilities imposed on providers. This recommendation describes an overall screening pipeline that would help relevant officials establish the framework for the synthetic nucleic acid screening required by Executive Order 14110 section 4.4(b)(i).

We first describe the broad outlines of the data to be collected and criteria to be checked at each step of the screening process. Subsequent recommendations describe in greater detail the components of this proposed screening workflow, including customer screening, communication with law enforcement, conformity assessment to verify that providers have effectively implemented the screening protocol, and the establishment of an external database for information about synthesis orders. Figure 4.1 shows a schematic view of the proposed screening protocol. The proposed screening workflow produces one of three possible outcomes: order fulfillment, order rejection, or order rejection with reporting to law enforcement.

Components of the screening process that are consistent with this and subsequent recommendations could be outsourced by providers to third-party commercial services. For exam-

FIGURE 4.1
Proposed Screening Framework

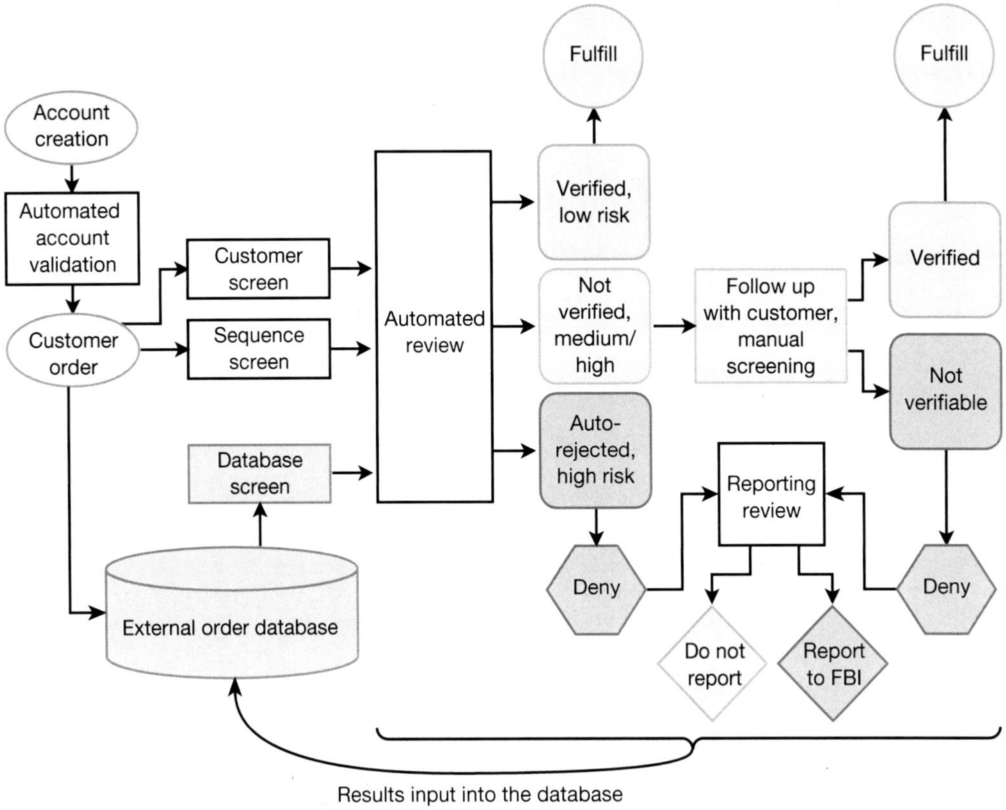

NOTE: The risk categories that result from the automated review are color-coded as follows: green = an order is classified as low risk; yellow = an order is classified as medium/high risk; red = an order is classified as high risk.

ple, some components of customer screening could be effectively conducted by third-party identity verification services, as we discuss in more detail below. Technical sequence screening could be conducted by third-party screening services (Aclid, undated; Battelle, undated; Raytheon BBN, undated; SecureDNA, undated).

The workflow begins with customer account creation on the provider's online order platform. Customers should be required to submit information that is sufficient for the provider to verify the customer's identity and establish that the customer has a legitimate purpose for the sequence(s) they order (Alexanian and Carter, 2024). In accordance with definitions given by the 2023 ASPR guidance, the term *customer* refers to the individual or entity that orders or requests synthetic nucleic acids from a provider (ASPR, 2023b). A *principal user* is an individual who makes an order for synthetic nucleic acids or oversees their use in the laboratory. An *end user* is an individual who possesses and uses the synthetic nucleic acids that they have received from a customer, principal user, or another end user. An order can have more than one principal user or more than one end user. In the recommendations that follow, we use the general term *customer screening* to collectively describe recommendations for screening the customer organization and the users of sequence products. Providers may reduce up-front costs by screening customers and users only after order submission but prior to confirming the order.

At the time of customer order creation, the customer submits a nucleic acid sequence for synthesis, along with information about how they will use it, how they will pay for it, and where the order should be shipped. The American Society of Gene and Cell Therapy (Barrett, 2022, p. 3) notes that the burden on synthetic oligonucleotide providers can be reduced if "customers preemptively provide information to verify their legitimacy when ordering synthetic oligonucleotides that they know contain sequences of concern." If the customer indicates the presence of an SOC, including a select agent or other controlled organism, the customer must be able to demonstrate a legitimate reason, including legal authorization if necessary, to possess and work with the sequence and any organisms that might result from its use. The provider order platform should prompt the customer to explain the intended use for the synthesized sequence and indicate whether the order contains a potentially dangerous sequence, including a select agent or other SOC. The customer should be asked to disclose whether the sequence or organism will be transferred to another party and the identity and location of that party. The identity and contact information for the principal and end users must be disclosed and verified.

To facilitate provider adoption and compliance, orders are grouped into risk categories through a process that can be almost entirely automated. The process takes the result of customer, sequence, and database screens (described below) as input, and it outputs one of the following three risk categories indicating whether the order: can be fulfilled immediately (low risk), requires further review (medium/high risk), or should be rejected immediately (high risk).

The risk categories are color-coded in Figure 4.1 as follows:

- **Green.** An order is classified as low risk if sequence screening returns no indications that the sequence is potentially dangerous, the customer's identity is verified, the provider can legally fulfill the order, and a database check determines that the order is not related to other concerning orders. An order that meets these criteria can be fulfilled without further review.
- **Yellow.** An unverified or medium/high risk order has at least one of the following characteristics: The customer is unverified, the sequence screening indicates that the sequence might be dangerous, the order sequence function and intended use are inconsistent, or a database check flags other possibly related orders that are individually or collectively suspicious. Yellow-flagged orders are subjected to follow-up screening, described below.
- **Red.** An order is classified as high risk or prohibited and automatically rejected if the customer, principal user, end user, or officer of the company is on a sanctions or prohibited customer list, if the order violates the provider's terms of service, or if fulfilling the order would violate any law.

Most orders should fall into the green category. Yellow orders are subjected to follow-up screening using a more-intensive process that involves gathering additional information about the identity and legitimacy of the customer, clarifying the intended use of the sequence, and verifying that the customer can legally and safely handle the synthesized sequence and any organism that might result from its use.

This biosecurity review step could be entirely or mostly automated from the provider perspective. The results of each screening component—customer, sequence, and database—should have a standardized format accepted as input into the review. The color-coded output categories have concrete and specific criteria that are amenable to automated checking or outsourcing to a service provider.

In our recommended screening workflow, the order must be rejected after follow-up screening if any of the following criteria are met:

- The customer remains unverified.
- The sequence itself is identified as potentially dangerous AND
  - the customer does not have a legitimate use for the sequence OR
  - the provider does not have authorization to produce or transfer the sequence or resulting agent.
- There is evidence that the customer has attempted to deliberately evade screening, such as by obfuscating or splitting across orders.
- The sequence order would be illegal to fulfill because the customer does not have authorization to use the sequence or fulfilling the order would violate the provider's terms of service.

The order must be rejected *and reported to law enforcement* if the sequence itself is identified as potentially dangerous via technical screening AND any of the following criteria are met:

- The customer is unverified, does not have authorization to use the sequence, or does not have a legitimate use for the sequence.
- The customer has attempted to deliberately evade screening.

When a rejected order is reported to law enforcement, the provider should communicate all order and customer information and all information collected during the automated and follow-up screening process. Providers may, at their discretion, reject a suspicious order or report an order to law enforcement even if these criteria are not met. A template report to law enforcement is given in Appendix A.

Implementation of this screening workflow and the more-specific recommendations described below may present opportunities to learn and refine screening, reporting, and conformity assessment processes. Best practices for these steps should be identified from provider and customer experiences and shared with industry stakeholders. Analysis of the generalizability of the screening methods to other dual use biotechnologies, such as the purchase of benchtop nucleic acid synthesizers, should be performed.

Establish a Third-Party Database to Log Information About Synthetic Nucleic Acid Orders and Periodically Check Records for Evidence of Suspicious Activity

In our review of the literature, we found that providers currently do not have a way of detecting split orders. We therefore recommend the establishment of a database system to record certain information about gene synthesis orders and customer information, administered by the government or an entity created or identified for this purpose. This entity would be responsible for logging information about synthetic gene orders and periodically checking records for evidence of suspicious activity. For each order, the provider should submit the information collected about the customer and some information about the order to the database, as described below.

Customers and providers may be wary of submitting sequence information to a third-party database because the sequence may constitute intellectual property or a trade secret. We describe the following three options for the submission of order information to the database:

- Submit all sequence information for every order.
- Submit sequence information only if the order is classified as "yellow" and subjected to secondary screening or if the order is ultimately rejected.
- Submit sequence information to the database only if the order is rejected.

In each case, customer and order data, as well as sequence information when appropriate, should be securely transmitted by the provider to the database, encrypted, and stored securely in the database (Alexanian and Nevo, 2024). Once the screening process has been completed for an order, the final disposition of that order (including data collected during any follow-up screening) should be reported to the database, along with the amount of sequence information appropriate for an order with that disposition. Provider terms of use should specify the circumstances under which the provider will share order and sequence information with the database or any third party.

The database should link records from the same customer (organization and principal user) using persistent identifiers generated from government-issued identification documents verified to the standards determined by the NIST risk assessment process described in subsequent recommendations. On submission of customer and order information, the database system should search for other orders by the same customer for evidence of identity fraud, evidence of splitting orders of potentially dangerous sequences across providers, or prior rejected orders. The database system should return to the provider information about previously denied or flagged orders from the same or other providers, as well as information about possible order splitting across providers. The database would not return to the provider any sequence information from other orders. Data retention policies for the database system should be established so that order history over at least eight years is present, following current 2023 ASPR guidance (ASPR, 2023b, p. 10).

Information gathered by providers for the purpose of customer verification, such as records of government-issued identification, should be stored in a privacy-preserving way and periodically screened using up-to-date fraud and forgery detection tools. When a previous order is determined to have been made with a fake or stolen identity document, information regarding that order and others associated with those identifying documents should be shared with law enforcement. In addition to checks conducted as part of order screening, the entity administering the database system should periodically run checks across all orders to detect evidence of split orders for dangerous sequences or other malicious intent. Unique identifiers derived from identification documents and supporting customer information, such as delivery addresses and browser fingerprints, should be used to detect orders from the same customer split across providers. Similarly, as new SOCs are added to technical screening systems and new analytical techniques are developed to detect split orders (Kane and Parker, 2024), retroactive screening should be run periodically by providers on historical order data and by the database system on historical order entries containing sequence data. Such a database would also support enhanced detection of unsanctioned reconnaissance efforts by malicious actors using multiple provider services to systematically identify and exploit vulnerabilities in the screening system.

Administration of the database system by a third party would reduce the compliance burden on providers because they would not be responsible for the retroactive screening of past orders. In addition, providers would not have direct access to the database and could not inspect orders submitted by any customer to other providers. Instead, the database system would perform these checks independently, returning to the provider only the information they need to know in order to complete the screening process for a given order. The overall purposes of the database would be the routine checking of new orders to detect split orders of potentially dangerous sequences to evade screening, periodic checking of previous orders, and facilitating the investigation of order history by law enforcement agencies when necessary.

Establish Customer Screening Requirements, Including Identity Verification Measures

We found that stronger customer screening is a promising way of keeping unauthorized or prohibited customers from ordering synthetic nucleic acid sequences. We recommend that providers verify the legitimacy of customer organizations and the identity of principal and end users of ordered sequences using standards to be set through a formal risk assessment process. Identity verification of individuals could be conducted periodically: for example, once when the individual is registered on the provider order platform and annually thereafter. NIST Special Publication 800-63-3 specifies relevant standards for three components of what is generally referred to as identity verification (Grassi, Garcia, and Fenton, 2020). These components are identity proofing, authentication, and (where appropriate) federation assurance. *Identity proofing* is the process of collecting and verifying identifying information (e.g., a government-issued identity document) about a person. *Authentication* means linking the identity information to the account or order being created (e.g., logging in to a website or using multifactor authentication). *Federation* enables these steps to be conducted by third parties.

Standards to prove the identity of an individual are set by NIST identity assurance levels (IALs), with level 1 being the least rigorous. Similarly, NIST defines authenticator assurance levels (AALs) for linking an identity to an account, and federation assertion levels specify standards for sharing identity verification information between the two parties engaged in the verification process and a third-party identification provider. Federation assurance is not always required but could allow for the centralization of identity proofing and authentication in a way that reduces the customer and provider compliance burden. Customer screening requirements would by default apply to all orders, however, differentiated IALs based on a NIST risk assessment and the prioritization of automated checks should be used to minimize the compliance burden.

Conduct a Risk Assessment to Set Standards for Customer Verification

We recommend that the U.S. federal government conduct a risk assessment to determine the risk levels of different synthetic nucleic acid user types. A defined process for this assessment is described in NIST Special Publication 800-63-3 (Grassi, Garcia, and Fenton, 2020) and the NIST Risk Management Framework (NIST, 2024). This process assesses the impact of failure of different identity verification processes for the industry and informs the choice of appropriate NIST standards of evidence required for identity proofing, authentication, and federation. The industry and customer compliance burden should also be considered when determining these standards.

We recommend that independent risk assessments be performed for four user groups across customer institution types: principal users of orders known to contain SOCs, end users of orders known to contain SOCs, principal users of orders not known to contain SOCs, and end users of orders not known to contain SOCs. The risk assessment should also depend on sequence risk categories. For example, any customer ordering components of the 1918 influenza virus sequence should be subject to a higher level of identity assurance. Edison and Esvelt (2024) propose a useful breakdown of customer risk groups by customer institution type and risk of DNA product. To minimize the compliance burden for providers and users, the risk assessment process should also define a risk level below which IAL1 is sufficient for identity assurance. At this assurance level, self-assertion of identity is accepted, and, therefore, compliance costs to both customers and providers would be minimal.

The risk assessment process should also determine the frequency and periodic or irregular interval at which identity proofing must take place. We recommend that a government entity be identified with the capability to carry out these assessments and that its results be used to set assurance levels for proofing, authentication, and federation for each user group.

IALs and AALs cover a spectrum of security and compliance burden settings. IAL3 and AAL3 require the physical presence of the individual for proofing and the possession of a hardware authenticator, respectively. Such measures may be appropriate for some nucleic acid synthesis customer groups, such as those authorized to place orders for biosafety level 4 labs. In contrast, IAL1 does not require external evidence. For this reason, different standards are likely to be appropriate for different user groups. Section 5.4 of NIST Special Publication 800-63-3 on the risk assessment process specifies that these assessments should be updated continuously to account for changes in the threat landscape (Grassi, Garcia, and Fenton, 2020, p. 22). This is particularly important in the context of nucleic acid synthesis, given the speed of development and variety of applications of the technology.

Customer and User Verification

Providers should collect information that is sufficient to identify the person creating the account and the principal and end users of the sequences associated with each order, up to the appropriate IAL determined through the risk assessment process. We describe an example of a NIST-compliant process based on one described in section 4.1 of NIST Special Publication 800-63A for identity proofing in the context of nucleic acid synthesis customer screening (Grassi et al., 2020, p. 5). First, the provider (or the identity proofing service acting on behalf of the provider) collects photos of the customer's driver's license and passport via a laptop computer camera. The provider validates the information supplied by checking a government registry, determines that the information supplied by the customer matches the registry records, and checks the document images for evidence of alterations. The customer provides a photo of themself, and the platform determines whether the person in the photo is the same person as the one shown on the license and passport. The provider validates the phone number of the customer by sending a code, which is then input into the provider's platform to confirm that the customer is in control of the phone number.

Once the provider has proofed the identity of a customer or user, that identity must be linked to the account through a process called *authentication*. An *authenticator* is something that the customer possesses and controls and can be used to authenticate their identity, such as passwords, hardware tokens, and fingerprints. NIST's AALs specify how this linking occurs:

> AAL2 provides high confidence that the claimant controls authenticator(s) bound to the subscriber's account. Proof of possession and control of two distinct authentication factors is required through secure authentication protocol(s). Approved cryptographic techniques are required at AAL2 and above. (Grassi, Garcia, and Fenton, 2020, p. vii)

Higher standards require such authentication processes as hardware-based multifactor authentication that increase confidence that the person accessing the service is the same as the subject of the identity proofing process or at least controls the authenticators linked to that identity. For example, this step could be completed via the submission of a password and an authenticator app code connected with a principal user identity that has been proofed to IAL2 standards within the last 18 months. When these steps—identity proofing and authentication at the chosen IAL and AAL—are successfully completed, we say that the customer has been *verified*. Identity proofing need not occur for each order, but authentication must occur for each order and each user of the order.

We emphasize that identity proofing customers and users may occur only occasionally. Authentication—for example, linking a user's identity to an account on the provider's order platform by logging in—might occur whenever an order is submitted. Identity proofing should expire at appropriate intervals, and authorization should end when a user's employment with a customer organization ends.

Facility Certifications

In addition to verifying customer identities, depending on the sequences being ordered, the provider should collect information about laboratory safety certifications and authorization to handle the sequences and the resulting agents or organisms.

Browser Information

In particular, the provider should collect information about the public IP address of the device creating the account, time and date of account creation, customer legal name, email address, phone number, shipping address, and payment method. In addition, the provider's online registration system should also collect the *browser fingerprint*, a collection of characteristics transmitted by the user's internet browser, at the time of account creation. Collected browser fingerprint information should include user-agent, accept, connection, content encoding, content language, list of plugins, cookies enabled, use of local/session storage, time zone, screen resolution and color depth, the list of fonts, list of hypertext transfer protocol headers, platform, Do Not Track browser setting, canvas, web graphics library vendor, web graphics library renderer, and use of an ad blocker (Alaca and van Oorschot, 2016; Laperdrix et al., 2020). Browser fingerprints associated with devices used to place orders should be stored as part of the customer screening information used to identify potential split orders in the third-party database. Collection of this information can be achieved through the use of automated third-party tools commonly used by the finance industry to counter payment and identity fraud and to support compliance with know-your-customer regulations (Fingerprint, undated; SHIELD, undated). These tools are useful in that context because they can be used to identify evidence that a user interacting with an online banking system is not who they claim to be or find patterns of suspicious transactions across users.

Sanctions Lists and Identity Obfuscation

The following automated customer screening checks are recommended. They are primarily developed from customer screening processes commonly used in the financial services industry and adapted for biotechnology products and services sold via e-commerce websites.

- Check the customer's legal and (where appropriate) business name against lists of prohibited individuals or organizations. Commercial sanctions screening software typically checks hundreds of lists (Moody's, undated; Thompson Reuters, undated-b), which include the U.S. Consolidated Screening List; United Nations sanctions list; U.S. Office of Foreign Assets Control Specially Designated Nationals and Blocked Persons List; U.S. Department of State Debarred Parties List; U.S. Bureau of Industry and Security Denied Persons List, Entity List, and Unverified List; UK Office of Financial Sanctions Implementation sanctions list; Consolidated List of European Union Financial Sanctions; Australian Government Department of Foreign Affairs and Trade Consolidated List; Consolidated Canadian Autonomous Sanctions List; United Nations Security Council Consolidated List; rexx systems sanctions screening software; and commercially available adverse media screening tools, such as Reuters CLEAR adverse media checks (Thompson Reuters, undated-a).
- Verify the customer's phone number and email via link or code.
- Check for the use of identity obfuscation systems.
  - Verify that the shipping address is not a post office box, mail forwarding service, or other disallowed address type. Providers should use automated tools (Experian, undated; Trulioo, undated) to conduct these checks but should retain the ability to overrule those tools when the results of the checks are ambiguous.

- Check the domain of the email address provided against a database of temporary and masked address domains.
- Check phone numbers provided against a register of voice-over-IP (VOIP) and other sets of numbers commonly used to evade phone verification systems.
- Check the IP address against the current list of Tor exit nodes and virtual private network (VPN) providers.
- Check that the customer's payment method is compliant with financial know-your-customer regulations.
- Check to determine if the domain listed on the organization website matches the domain of the email address provided by the customer.

Many of these automated checks are not definitive but do provide useful context as indicators of risk or as inputs for follow-up screening and inclusion in reports to law enforcement. A customer order should be rejected if any of the following occur:

- placement of an order via the Tor browser or over the Tor network
- attempted use of an anonymity-focused cryptocurrency or cash
- use of a temporary email address or VOIP number
- detection of a stolen or fraudulent government-issued identity document
- a match with a sanctions or watch list for which the provider is unable to determine is in error.

Flags that may indicate risk but can be overridden by a provider include the use of a masked, personal, or noninstitutional email address or credit card, the use of a VPN, an ambiguous address type, or a match with a watch list that on further investigation by the provider is confirmed to be a different person. Once these checks have been carried out and the customer has been verified, the customer is considered verified and prescreened for the duration of the order screening process.

Expand the Scope of Customer Screening for All Orders Regardless of Whether They Match a Sequence of Concern

The *Framework for Nucleic Acid Synthesis Screening* from the Fast Track Action Committee on Synthetic Nucleic Acid Procurement Screening (2024) specifies that compliant providers must screen customers whose order contains an SOC. We recommend that this framework be updated to require that all customers be subject to screening, regardless of whether the order contains an SOC. Providers should conduct a basic set of automated customer screening checks on all customers, including verifying that a customer is not using the Tor network to obfuscate their identity when ordering nucleic acid products and services and checking whether the customer appears on a sanctions or adverse media watch list.

Burden of Customer Screening for Providers

Rigorous customer screening may impose burdens on both providers and customers. Writing about providers' perception of the cost of sequence screening, Kane and Parker (2024, p. 5) note that "the largest cost was in time, as the review of flagged sequences by humans slows down the screening process." Automated customer screening methods may be particularly important as the price of nucleic acid synthesis decreases and bottlenecks—such as human review—become a proportionally larger part of the total cost. Kane and Parker (2024, p. 6) also note that "[a]utomation . . . also reduces the cost of internal pipelines, as this reduces the time and financial cost of human review." We recommend that providers outsource customer screening when possible to established, certified, commercially available tools. Third-party screening services can also reduce barriers to entry and competition, capture economies of scale, and allow firms to focus on developing their technology rather than customer screening. Carter (2024, p. 21) writes that

> [t]he advantage of third-party customer screening is that it can ensure consistent customer screening practices across the community and can provide some efficiency by enabling customers to become verified only once rather than by each vendor separately.

Third-party screening would also help large institutional customers, such as bioscience companies and universities, adhere to customer screening rules. These large institutions may place many synthetic DNA orders for use in laboratories (by principal and end users) across their organizations. Third-party screening could preapprove these users so that identity verification would not have to be repeated for every provider.

Establish Criteria and Data Requirements for Sharing Information About Suspicious Orders with Law Enforcement Agencies

We have found that establishing clear criteria for reporting suspicious orders to law enforcement would reduce ambiguity for providers and improve communication about concerning or dangerous orders. In the proposed screening workflow, if the order is to be rejected (not fulfilled) by the provider, a second decision must be made: whether to report the customer and order to a relevant law enforcement agency. The provider's screening assessments provide the evidence to be used in the following sufficient criteria for reporting:

- The customer, the specific order from that customer, and historical orders from the customer **must** be reported to authorities if the sequence itself is identified as potentially dangerous via technical screening AND follow-up screening has established that
  - there is evidence that the customer has attempted to deliberately evade screening, such as by obfuscating or splitting across orders, OR

- there is evidence that indicates the sequence order would be illegal to fulfill or violates the provider's terms of service, OR
- the customer identity is not proofed, the customer refuses to provide the required certifications to handle the sequence, or the customer does not have a legitimate use for the sequence.

The provider may still consider reporting or communicating with law enforcement about a suspicious order at any stage in the screening process, even if the above criteria are not met. Beyond the sufficient criteria for reporting described above, providers should maintain communication with relevant enforcement bodies, including the local FBI WMD coordinator. In accordance with ASPR and IGSC guidelines, providers should contact the WMD coordinator with informal questions and concerns.

When a decision is made to reject an order and report it to law enforcement, the provider should supply the relevant authorities with a standardized document containing all information collected about the customer, the order, the information gained from the screening steps, the assessments made during the initial biosecurity review and follow-up screening, and any additional correspondence with customer representatives. A template report to law enforcement is presented in Appendix A. Provider terms of service should make clear that providers may share order sequence data or other order or customer information with law enforcement under the circumstances described above.

Implement Procedures and Standards for Assessing the Conformity of Providers' Screening Practices with Requirements

To ensure that gene synthesis providers comply with the screening protocol, we recommend implementing procedures and standards for assessing the conformity of providers' screening practices with requirements. The U.S. government currently has no way of assessing providers' conformity with a screening framework. Executive Order 14110 links provider conformity with requirements to customers' ability to receive federal funds, so the U.S. government must be able to verify that providers are conformant. The conformity assessment process described here is designed to minimize additional burden on providers beyond that of compliance with requirements. We recommend a conformity assessment composed of two components: routine synthesis screening audits and red-teaming exercises. These components are augmented by recordkeeping of provider compliance status and providing a public point of contact for reporting vulnerabilities in screening.

Synthesis screening audits should be conducted by a third party using a formalized screening protocol assessment process to assess providers' conformity with established screening rules. An example questionnaire appears in Appendix B. This questionnaire is designed to guide routine synthesis screening audits and help synthetic gene providers conduct self-evaluation to ensure that they are in compliance with the screening protocol described here. The *Framework for Nucleic Acid Synthesis Screening* (Fast Track Action Committee on Synthetic Nucleic Acid Procurement Screening, 2024, p. 4) requires providers to "[a]ttest to implementing this screening framework"; such an attestation could be based on the completion of a questionnaire, such as the one presented in Appendix B. The questionnaire is made up of three sections: provider information, customer screening procedures, and order screening procedures. The questionnaire reflects the screening protocol developed in this report and the 2023 ASPR guidelines and IGSC rules. The questionnaire should be regularly updated to reflect new threats and best practices for screening. Additionally, the government agency overseeing synthesis screening evaluations should work with providers of synthetic gene products to ensure that they are aware of the criteria, as well as other assessment procedures described below, and are implementing best practices to prevent the misuse of their products. The oversight agency should work with law enforcement agencies to ensure that law enforcement entities are aware of the criteria and are prepared to respond to any threats that are identified. The oversight agency should publish biosecurity scenarios, such as those presented in the ASPR companion guidance (2023a), and make the scenarios available to providers for use in evaluating their own screening processes. Violations or issues found during third-party synthesis screening audits should be communicated back to the provider and recorded by the oversight agency. Providers that accumulate enough violations to be considered to have failed the audit should be flagged as not conforming to the requirements.

Red-teaming exercises are designed to test the efficacy of the screening system in identifying orders that should be flagged under the guidelines and weaknesses in current protocols. Red teaming should be conducted by a nongovernment third party, in cooperation with law enforcement. The red team will attempt to actively evade the sequence and customer screening controls to see if it can successfully procure orders that those controls should block. One way for the red team to test customer screening protocols is to create accounts representing fake institutions or customer identities that appear legitimate but would fail the customer screening protocol described above. Red-teaming results will be shared confidentially with providers and used to update screening protocols and rules. Red team results are sensitive, and vulnerabilities identified during red teaming should not be publicized before those vulnerabilities are addressed by providers. However, lessons learned by the government and providers during red-teaming exercises should be shared with the industry to improve screening practices and to inform future requirements or regulations. Given that the submission of an apparently illegitimate order under the identity of an existing customer could have at least the appearance, if not the actual possibility, of putting that customer's reputation at risk, red teaming may not be a very good tool for assessing the risk of insider threats from established synthesis customers. Instead, its usefulness might be limited to evaluating provider procedures for validating new customers and screening their orders. Outside red-teaming exercises, the government should provide a public point of contact for reporting vulnerabilities in screening discovered by outside researchers.

The oversight agency should maintain an up-to-date list of providers that conform to regulations for the use of customers that receive federal funds. The list of conformant providers should not be made public to prevent malicious actors from inferring which providers do not have effective synthesis screening programs. However, providers should be permitted to advertise that they are in compliance and can conduct business with recipients of federal funds. To prevent nonconformant providers from falsely claiming to be conformant, the oversight agency should provide a secure way for customers to ensure that a given provider appears on that list. For example, customers could query a system operated by the oversight agency regarding the conformity status of a given provider, and the system could return a time-stamped approval, valid for a given time interval, if the provider is deemed to be conformant. Any government-funded customer would need to be able to demonstrate that all nucleic acid orders were placed within the window of validity of such a time-stamped approval. In addition, the government should create a mechanism for providers to appeal their exclusion from the list of approved providers.

Example Template Report to Law Enforcement

This appendix presents an example of a template report providers could send to law enforcement.

- Provider information
  - Provider name, address, phone number, website
  - Name and contact information of provider biosecurity officer making the report
  - Brief statement of provider policy and the nature of the report as follows:
    - [Provider name] requires that nucleic acid synthesis customers have a legitimate and verifiable peaceful use for the sequence(s) they order. We have determined that the following customer order meets our criteria for reporting to law enforcement. For more information about our screening process and reporting criteria, see . . .

- Customer information
  - Individual completing order, point of contact address, email, phone number
  - Account creation data
  - Company or organization name, address
  - Shipping information, receiver
  - Customer certifications

- Customer and communication metadata
  - Date
  - IP address
  - User information

- Order information
  - Nucleotide sequence
  - Vector
  - Intended use
  - Certifications, permission to handle or transfer sequence or resulting organism

- Results of initial screening
  - Customer screening results, protocol or service used
  - Sequence screening results, protocol used
  - External database search results
  - Initial assessment and reason for follow-up screening

- Results of follow-up screening
  - Additional information collected
  - Assessment of customer and order and reason for reporting the order to law enforcement
  - Order disposition: not fulfilled

- Copy of all communications with customer, including denial of order
- Customer's prior order information from this and possibly other providers

APPENDIX B

Example Nucleic Acid Synthesis Screening Audit Questionnaire

This appendix presents an example of a nucleic acid synthesis screening audit questionnaire. This questionnaire is designed to guide routine synthesis screening audits and help synthetic gene providers conduct self-evaluation for compliance.

Section 1: Provider Information

Name of gene synthesis provider:
Address:
Phone number:
Email:
Date of last nucleic acid synthesis screening audit:
Is your organization an approved provider for federally funded research projects? Yes or no
Point of contact for reporting vulnerabilities in screening discovered by external researchers:

Section 2: Customer Screening Procedures

1. Does the provider use a certified identity collection and verification system or systems that satisfy standards determined by the NIST risk assessment process for both principal and end users?

2. Does the provider use a conformant sequence screening tool to check all orders for SOCs prior to order fulfillment?

3. Does the provider check the customer's legal and (where appropriate) business name against lists of prohibited individuals or organizations? Check all that apply.
   - United Nations sanctions list
   - U.S. Consolidated Screening List
   - U.S. Office of Foreign Assets Control Specially Designated Nationals List and Blocked Persons List
   - U.S. Department of State Debarred Parties List

- U.S. Bureau of Industry and Security Denied Persons List, Entity List, and Unverified List
- UK Office of Financial Sanctions Implementation sanctions list
- Consolidated List of European Union Financial Sanctions
- Australian Government Department of Foreign Affairs and Trade Consolidated List
- Consolidated Canadian Autonomous Sanctions List
- United Nations Security Council Consolidated List
- rexx systems sanctions screening software.

4. Does the provider verify the customer phone number and email via link or code?

5. Does the provider check for use of identity obfuscation systems?
   - Does the provider verify that the shipping address is not a post office box, mail forwarding service, or other disallowed address type?
   - Does the provider check the domain of the email address provided against a database of temporary and masked address domains?
   - Does the provider check phone numbers provided against a register of VOIP and other sets of numbers commonly used to evade phone verification systems?
   - Does the provider check the IP address against the current list of Tor exit nodes and VPN providers?
   - Does the provider check that the customer's payment method is compliant with financial know-your-customer regulations?
   - Does the provider check to determine if the domain listed on the organization website matches the domain of the email address provided by the customer?

Section 3: Order Screening

6. Does the provider order platform prompt the customer to explain the intended use for the synthesized sequence and indicate whether the order contains a potentially dangerous sequence, including a select agent or other SOC?

7. Does the provider have a clear mechanism for the customer to provide proof of authorization to possess and work with the sequence and any organisms that might result from its use if the order contains a potentially dangerous sequence, including a select agent or other SOC?

8. Does the provider have a clear mechanism for the customer to disclose whether the sequence or organism will be transferred to another party and to submit the identity and location of that party?

9. If the order is categorized as yellow, does the provider conduct follow-up screening before confirming or fulfilling the order, including gathering or confirming all the information listed in the decision criteria below?

10. Is the order rejected after follow-up screening if any of the following criteria are met: the customer remains unverified, the sequence itself is identified as potentially dangerous and the customer does not have a legitimate use for the sequence or authorization to handle the sequence or resulting agent, there is evidence that the customer has attempted to deliberately evade screening, or the sequence order would be illegal to fulfill or would violate the provider's terms of service?

11. Is the order rejected and reported to law enforcement if the sequence itself is identified as potentially dangerous via technical screening and any of the following criteria are met: the customer is unverified, does not have authorization to use the sequence, or does not have a legitimate use for the sequence; the customer has attempted to deliberately evade screening; or the sequence order would be illegal to fulfill?

12. Is information about the customer and order submitted to the third-party database system at the time of order creation?

13. Is the sequence information submitted to the database in accordance with guidelines?

14. Is order and screening information stored for a period of eight years?

Section 4: Communication with Law Enforcement

15. Are the following data points collected and saved by the provider, such that they are available to law enforcement officials when required?
    - Provider
        - Company name, address
        - Provider identification
        - Biosecurity officer, reporting employee
    - Principal and end users
        - Individual completing order, point of contact address, email, phone number
        - Account creation data
        - Company or organization name, address
        - Shipping information, receiver
        - Stated use
        - Customer history of use, intentions

- Order metadata
  - Date
  - IP address
  - User information
  - Reason for ordering, use
- Order sequence
  - Nucleotide sequence
  - Vector
  - Sequence metadata
- Reason for flag
  - Customer screening results, protocol or service used
  - Sequence screening results, protocol used
  - External database flag
  - Follow-up screening results
- Copy of all communications with customer, including denial of order
- Customer's prior order information from this and possibly other providers.

Abbreviations

AAL	authenticator assurance level
AI	artificial intelligence
ASPR	Administration for Strategic Preparedness and Response
DNA	deoxyribonucleic acid
dsDNA	double-stranded DNA
EBRC	Engineering Biology Research Consortium
FBI	Federal Bureau of Investigation
FSAP	Federal Select Agent Program
HHS	U.S. Department of Health and Human Services
IAL	identity assurance level
IASB	International Association for Synthetic Biology
IBBIS	International Biosecurity and Biosafety Initiative for Science
IGSC	International Gene Synthesis Consortium
IP	internet protocol
NCBI	National Center for Biotechnology Information
NIH	National Institutes of Health
NIST	National Institute of Standards and Technology
NTI	Nuclear Threat Initiative
OSTP	White House Office of Science and Technology Policy
RNA	ribonucleic acid
SOC	sequence of concern
Tor	The Onion Router
UK	United Kingdom
VOIP	voice over internet protocol
VPN	virtual private network
WHO	World Health Organization
WMD	weapons of mass destruction

References

Aclid, homepage, undated. As of June 10, 2024:
https://www.aclid.bio

Administration for Strategic Preparedness and Response, "Companion Guide to Assist in Implementing the Recommendations of the Screening Framework Guidance for Providers and Users of Synthetic Nucleic Acids," U.S. Department of Health and Human Services, October 2023a. As of June 10, 2024:
https://aspr.hhs.gov/legal/synna/Documents/SynNA-Companion-Guide-508.pdf

Administration for Strategic Preparedness and Response, "Screening Framework Guidance for Providers and Users of Synthetic Nucleic Acids," U.S. Department of Health and Human Services, October 2023b. As of June 10, 2024:
https://aspr.hhs.gov/legal/synna/Documents/SynNA-Guidance-2023.pdf

Agarwal, Shruti, Hany Farid, Tarek El-Gaaly, and Ser-Nam Lim, "Detecting Deep-Fake Videos from Appearance and Behavior," *2020 IEEE International Workshop on Information Forensics and Security*, December 6–11 2020.

Alaca, Furkan, and P. C. van Oorschot, "Device Fingerprinting for Augmenting Web Authentication: Classification and Analysis of Methods," *Proceedings of the 32nd Annual Conference on Computer Security Applications*, December 2016.

Aldhous, Peter, "The Bioweapon Is in the Post," *New Scientist*, No. 2525, November 2005.

Alexanian, Tessa, and Sarah R. Carter, "Verifying Legitimacy: Findings from the Customer Screening Working Group, 2020–2023," white paper, International Biosecurity and Biosafety Initiative for Science, February 2024.

Alexanian, Tessa, and Sella Nevo, "Supporting Follow-Up Screening for Flagged Nucleic Acid Synthesis Orders," Council on Strategic Risks, Janne E. Nolan Center on Strategic Weapons, *Briefer*, No. 60, May 7, 2024.

ASPR—*See* Administration for Strategic Preparedness and Response.

Assembly Bill No. 70, Gene Synthesis Providers, California Legislature—2021–2022 Regular Session, 2021.

Assembly Bill No. 1963, Chapter 5.7. Gene Synthesis Providers, California Legislature—2021–2022 Regular Session, August 26, 2022.

Balaji, Advait, Bryce Kille, Anthony D. Kappell, Gene D. Godbold, Madeline Diep, R. A. Leo Elworth, Zhiqin Qian, Dreycey Albin, Daniel J. Nasko, Nidhi Shah, Mihai Pop, Santiago Segarra, Krista L. Ternus, and Todd J. Treangen, "SeqScreen: Accurate and Sensitive Functional Screening of Pathogenic Sequences via Ensemble Learning," *Genome Biology*, Vol. 23, No. 1, June 2022.

Barrett, David, "Re: ASGCT's Comments on Proposed Revisions to the Federal Screening Framework Guidance for Providers of Synthetic Double-Stranded DNA," letter to the U.S. Department of Health and Human Services, American Society of Gene and Cell Therapy June 29, 2022. As of June 10, 2024:
https://asgct.org/global/documents/asgct-oligonucleotide-comments-final.aspx

Batalis, Stephanie, Caroline Schuerger, Gigi Kwik Gronvall, and Matthew E. Walsh, "Safeguarding Mail-Order DNA Synthesis in the Age of Artificial Intelligence," *Applied Biosafety*, Vol. 29, No. 2, 2024.

Battelle, "UltraSEQ™ Advanced DNA Sequencing Threat Analysis," webpage, undated. As of June 10, 2024:
https://www.battelle.org/markets/health/chemical-and-biological-countermeasures/biosecurity-pandemic-preparedness/threatseq

Bügl, Hans, John P. Danner, Robert J. Molinari, John T. Mulligan, Han-Oh Park, Bas Reichert, David A. Roth, Ralf Wagner, Bruce Budowle, Robert M. Scripp, Jenifer A. L. Smith, Scott J. Steele, George Church, and Drew Endy, "DNA Synthesis and Biological Security," *Nature Biotechnology*, Vol. 25, No. 6, June 2007.

Bureau of Industry and Security, "Commerce Control List (CCL)," webpage, U.S. Department of Commerce, undated. As of June 10, 2024:
https://www.bis.doc.gov/index.php/regulations/commerce-control-list-ccl

Cabinet Office, "Biological Security Strategy: Summary of Public Response," webpage, UK Government, June 15, 2022. As of June 10, 2024:
https://www.gov.uk/government/calls-for-evidence/biological-security-strategy-call-for-evidence/outcome/biological-security-strategy-summary-of-public-response

Carlson, Rob, "DNA Cost and Productivity Data, aka 'Carlson Curves,'" *synthesis* blog, October 26, 2022.

Carter, Sarah R., *Developing a Customer Screening Framework for the Life Sciences*, Blueprint Biosecurity, March 2024.

Carter, Sarah R., and Robert M. Friedman, *DNA Synthesis and Biosecurity: Lessons Learned and Options for the Future*, J. Craig Venter Institute, October 2015.

Carter, Sarah R., Jaime M. Yassif, and Christopher R. Isaac, "*Benchtop DNA Synthesis Devices: Capabilities, Biosecurity Implications, and Governance*," Nuclear Threat Initiative, May 2023.

Casagrande, Rocco, Jennifer Byers, Nick Wills, Skye Wills, Yu-Ming Liou, Alicia Isom, and Elizabeth Kramer, *The Custom Synthetic Nucleic Acid Industry and Biosecurity: A Systems Analysis*, Gryphon Scientific, December 2007. As of June 10, 2024:
http://gryphonsci.wpengine.com/wp-content/uploads/2019/01/DNA-Synthetic-Nucleic-Acid-Industry-Analysis_DHSGryphon_cleared.pdf

Cello, Jeronimo, Aniko V. Paul, and Eckard Wimmer, "Chemical Synthesis of Poliovirus cDNA: Generation of Infectious Virus in the Absence of Natural Template," *Science*, Vol. 297, No. 5583, July 2002.

Check Hayden, Erika, "Keeping Genes Out of Terrorists' Hands," *Nature*, Vol. 461, No. 7260, September 2009.

Church, George, "Let Us Go Forth and Safely Multiply," *Nature*, Vol. 438, No. 7067, November 2005.

Cybersecurity and Infrastructure Security Agency, "CISA Red Team Shares Key Findings to Improve Monitoring and Hardening of Networks," cybersecurity advisory, Alert Code AA23-059A, February 28, 2023.

DiEuliis, Diane, Sarah R. Carter, and Gigi Kwik Gronvall, "Options for Synthetic DNA Order Screening, Revisited," *mSphere*, Vol. 2, No. 4, August 2017.

Diggans, James, and Emily Leproust, "Next Steps for Access to Safe, Secure DNA Synthesis," *Frontiers in Bioengineering and Biotechnology*, Vol. 7, April 2019.

EBRC—*See* Engineering Biology Research Consortium.

Edison, Rey, and Kevin M. Esvelt, *DNA Synthesis Screening: A Technical and Policy Brief Relevant to E.O. 14110*, 2024. As of June 10, 2024:
https://drive.google.com/file/d/1h-rFDizahHC3c4tzChqVqkDq0I8Nbg0b/view

Engineering Biology Research Consortium, *Security Screening in Synthetic DNA Synthesis: Recommendations for Updated Federal Guidance*, April 2022.

Executive Order 14110, "Safe, Secure, and Trustworthy Development and Use of Artificial Intelligence," Executive Office of the President, October 30, 2023.

Experian, "Address Verification," webpage, undated. As of June 10, 2024:
https://www.edq.com/address-verification/

Falco, "What You Can Do with Falco Today," undated. As of June 10, 2024:
https://falco.org/about/use-cases/#threat-detection

Fast Track Action Committee on Synthetic Nucleic Acid Procurement Screening, *Framework for Nucleic Acid Synthesis Screening*, National Science and Technology Council, April 2024.

Federal Select Agent Program, "Security Risk Assessments," webpage, Centers for Disease Control and Prevention and Animal and Plant Health Inspection Service, January 11, 2022. As of June 10, 2024:
https://www.selectagents.gov/compliance/risk.htm

Federal Select Agent Program, homepage, Centers for Disease Control and Prevention and Animal and Plant Health Inspection Service, June 5, 2024. As of June 10, 2024:
https://www.selectagents.gov

Fingerprint, homepage, undated. As of June 10, 2024:
https://fingerprint.com

FSAP—*See* Federal Select Agent Program.

Garfinkel, Michele S., Drew Endy, Gerald L. Epstein, and Robert M. Friedman, "Synthetic Genomics: Options for Governance," *Industrial Biotechnology*, Vol. 3, No. 4, Winter 2007.

Gibson, Daniel G., Gwynedd A. Benders, Cynthia Andrews-Pfannkoch, Evgeniya A. Denisova, Holly Baden-Tillson, Jayshree Zaveri, Timothy B. Stockwell, Anushka Brownley, David W. Thomas, Mikkel A. Algire, Chuck Merryman, Lei Young, Vladimir N. Noskov, John I. Glass, J. Craig Venter, Clyde A. Hutchison III, and Hamilton O. Smith, "Complete Chemical Synthesis, Assembly, and Cloning of a *Mycoplasma genitalium* Genome," *Science*, Vol. 319, No. 5867, February 2008.

Gibson, Daniel G., John I. Glass, Carole Lartigue, Vladimir N. Noskov, Ray-Yuan Chuang, Mikkel A. Algire, Gwynedd A. Benders, Michael G. Montague, Li Ma, Monzia M. Moodie, Chuck Merryman, Sanjay Vashee, Radha Krishnakumar, Nacyra Assad-Garcia, Cynthia Andrews-Pfannkoch, Evgeniya A. Denisova, Lei Young, Zhi-Qing Qi, Thomas H. Segall-Shapiro, Christopher H. Calvey, Prashanth P. Parmar, Clyde A. Hutchison III, Hamilton O. Smith, and J. Craig Venter, "Creation of a Bacterial Cell Controlled by a Chemically Synthesized Genome," *Science*, Vol. 329, No. 5987, July 2010.

Gibson, Daniel G., Lei Young, Ray-Yuan Chuang, J. Craig Venter, Clyde A. Hutchison III, and Hamilton O. Smith, "Enzymatic Assembly of DNA Molecules Up to Several Hundred Kilobases," *Nature Methods*, Vol. 6, No. 5, May 2009.

Grassi, Paul A., James L. Fenton, Naomi B. Lefkovitz, Jamie M. Danker, Yee-Yin Choong, Kristen K. Greene, and Mary F. Theofanos, *Digital Identity Guidelines: Enrollment and Identity Proofing*, National Institute of Standards and Technology, NIST Special Publication 800-63A, June 2017, updated March 2, 2020.

Grassi, Paul A., Michael E. Garcia, and James L. Fenton, *Digital Identity Guidelines*, National Institute of Standards and Technology, NIST Special Publication 800-63-3, June 2017, updated March 2, 2020.

HHS—*See* U.S. Department of Health and Human Services.

Hoffmann, Stefan A., James Diggans, Douglas Densmore, Junbiao Dai, Tom Knight, Emily Leproust, Jef D. Boeke, Nicole Wheeler, and Yizhi Cai, "Safety by Design: Biosafety and Biosecurity in the Age of Synthetic Genomics," *iScience*, Vol. 26, No. 3, March 2023.

House Bill 1256, Maryland Department of Health—Gene Synthesis Providers and Manufacturers of Gene Synthesis Equipment—Certification, Maryland General Assembly, 2021.

IASB—*See* International Association for Synthetic Biology.

IBBIS—*See* International Biosecurity and Biosafety Initiative for Science.

ID Analyzer, "Supported Documents," undated. As of June 10, 2024:
https://www.idanalyzer.com/solutions/supported-documents.html

IGSC—*See* International Gene Synthesis Consortium.

Ingraham, John B., Max Baranov, Zak Costello, Karl W. Barber, Wujie Wang, Ahmed Ismail, Vincent Frappier, Dana M. Lord, Christopher Ng-Thow-Hing, Erik R. Van Vlack, Shan Tie, Vincent Xue, Sarah C. Cowles, Alan Leung, João V. Rodrigues, Claudio L. Morales-Perez, Alex M. Ayoub, Robin Green, Katherine Puentes, Frank Oplinger, Nishant V. Panwar, Fritz Obermeyer, Adam R. Root, Andrew L. Beam, Frank J. Poelwijk, and Gevorg Grigoryan, "Illuminating Protein Space with a Programmable Generative Model," *Nature*, Vol. 623, No. 7989, November 2023.

International Association for Synthetic Biology, "The IASB Code of Conduct for Best Practices in Gene Synthesis," November 3, 2009.

International Biosecurity and Biosafety Initiative for Science, homepage, undated. As of June 10, 2024:
https://ibbis.bio

International Gene Synthesis Consortium, "World's Top Gene Synthesis Companies Establish Tough Biosecurity Screening Protocol," press release, November 19, 2009.

International Gene Synthesis Consortium, "Harmonized Screening Protocol v2.0," November 19, 2017.

International Meeting on Synthetic Biology, "Declaration of the Second International Meeting on Synthetic Biology," revised public draft, May 29, 2006.

iProov, *Threat Intelligence Report 2024: The Impact of Generative AI on Remote Identity Verification*, 2024.

Jumio, "Identity Verification," webpage, undated. As of June 10, 2024:
https://www.jumio.com/products/identity-verification/

Kane, Arianne, and Michael T Parker, "Screening State of Play: The Biosecurity Practices of Synthetic DNA Providers," *Applied Biosafety*, Vol. 29, No. 2, 2024.

Kantara Initiative, homepage, undated. June 10, 2024:
https://kantarainitiative.org

Keesing Technologies, "Check ID Documents from Around the Globe," webpage, undated. As of June 10, 2024:
https://www.keesingtechnologies.com/documentchecker/id-documents/

Koblentz, Gregory D., "The *De Novo* Synthesis of Horsepox Virus: Implications for Biosecurity and Recommendations for Preventing the Reemergence of Smallpox," *Health Security*, Vol. 15, No. 6, November/December 2017.

Langenkamp, Max, "Clarifying the Problem of DNA Screening," Social Science Research Network, September 12, 2023.

Laperdrix, Pierre, Nataliia Bielova, Benoit Baudry, and Gildas Avoine, "Browser Fingerprinting: A Survey," *ACM Transactions on the Web*, Vol. 14, No. 2, April 2020.

Lewis, Nicholas C., "Inkshield," webpage, RepRap, November 15, 2015. As of June 10, 2024: https://reprap.org/wiki/Inkshield

LexisNexis Risk Solutions, "Watchlist Screening," webpage, undated. As of June 10, 2024: https://risk.lexisnexis.com/global/en/financial-services/financial-crime-compliance/watchlist-screening

Liu, Jianying, Longchao Zhu, Lingling Mei, Yuanyuan Liu, Yuanyuan Qu, Yulin Yuan, Mengjie Zhang, Fenfen Zhang, Yingyi Cao, Wanbo Tai, Jun Ma, Min Zheng, Xiaolu Shi, Yang Liu, and Gong Cheng, "An Efficient and Safe Trans-Complementation System of Mpox Virus Reproduces Authentic Viral Architecture and Infection," bioRxiv, 2023.

McLoughlin, Ian, "Secure Embedded Systems: The Threat of Reverse Engineering," *2008 14th IEEE International Conference on Parallel and Distributed Systems*, December 8–10, 2008.

Mihara, Tomoko, Yosuke Nishimura, Yugo Shimizu, Hiroki Nishiyama, Genki Yoshikawa, Hideya Uehara, Pascal Hingamp, Susumu Goto, and Hiroyuki Ogata, "Linking Virus Genomes with Host Taxonomy," *Viruses*, Vol. 8, No. 3, March 2016.

Millett, Piers, Tessa Alexanian, Kathryn R. Brink, Sarah R. Carter, James Diggans, Megan J. Palmer, Ryan Ritterson, Jonas B. Sandbrink, and Nicole E. Wheeler, "Beyond Biosecurity by Taxonomic Lists: Lessons, Challenges, and Opportunities," *Health Security*, Vol. 21, No. 6, November/December 2023.

Minshull, Jeremy, and Ralf Wagner, "Preventing the Misuse of Gene Synthesis," *Nature Biotechnology*, Vol. 27, No. 9, September 2009.

Monero, homepage, undated. As of June 10, 2024: https://www.getmonero.org

Moody's, "Sanctions Compliance," webpage, undated. As of June 10, 2024: https://www.moodys.com/web/en/us/kyc/solutions/sanctions.html

Moulin, Solène Lugon, Céline Weyermann, and Simon Baechler, "An Efficient Method to Detect Series of Fraudulent Identity Documents Based on Digitised Forensic Data," *Science & Justice*, Vol. 62, No. 5, September 2022.

National Academies of Sciences, Engineering, and Medicine, *Biodefense in the Age of Synthetic Biology*, National Academies Press, 2018.

National Institute of Standards and Technology, "NIST Risk Management Framework," webpage, updated May 8, 2024. As of June 10, 2024: https://csrc.nist.gov/projects/risk-management/about-rmf

National Library of Medicine, "NCBI Virus," webpage, undated. As of June 10, 2024: https://www.ncbi.nlm.nih.gov/labs/virus/vssi/#/

National Library of Medicine, "GenBank Overview," webpage, December 2022. As of June 10, 2024: https://www.ncbi.nlm.nih.gov/genbank/

National Science Advisory Board for Biosecurity, *Addressing Biosecurity Concerns Related to the Synthesis of Select Agents*, National Institutes of Health, December 2006.

National Security Commission on Emerging Biotechnology, "DNA 101: Reading, Writing, and Editing," February 2024a. As of June 10, 2024:
https://www.biotech.senate.gov/wp-content/uploads/2024/02/NSCEB_DNA.pdf

National Security Commission on Emerging Biotechnology, "Gene Synthesis Security," April 2024b. As of June 10, 2024:
https://www.biotech.senate.gov/press-releases/gene-synthesis-security/

Newsom, Gavin, letter to members of the California State Assembly regarding Assembly Bill 70, Office of the Governor, October 5, 2021. As of June 10, 2024:
https://www.gov.ca.gov/wp-content/uploads/2021/10/AB-70-PDF.pdf

Nguyen, Eric, Michael Poli, Matthew G. Durrant, Armin W. Thomas, Brian Kang, Jeremy Sullivan, Madelena Y. Ng, Ashley Lewis, Aman Patel, Aaron Lou, Stefano Ermon, Stephen A. Baccus, Tina Hernandez-Boussard, Christopher Ré, Patrick D. Hsu, and Brian L. Hie, "Sequence Modeling and Design from Molecular to Genome Scale with Evo," bioRxiv, 2024.

NIST—*See* National Institute of Standards and Technology.

Nouri, Ali, and Christopher F. Chyba, "Proliferation-Resistant Biotechnology: An Approach to Improve Biological Security," *Nature Biotechnology*, Vol. 27, No. 3, March 2009.

Noyce, Ryan S., Seth Lederman, and David H. Evans, "Construction of an Infectious Horsepox Virus Vaccine from Chemically Synthesized DNA Fragments," *PLoS ONE*, Vol. 13, No. 1, January 2018.

NTI—*See* Nuclear Threat Initiative.

Nuclear Threat Initiative, "Preventing the Misuse of DNA Synthesis Technology," webpage, undated. As of June 10, 2024:
https://www.nti.org/about/programs-projects/project/preventing-the-misuse-of-dna-synthesis-technology/

Proof, "Lock in Trust with Identity Verification for Every Customer," webpage, undated. As of June 10, 2024:
https://www.proof.com/product/proof-of-identity

Raytheon BBN, "Fast-NA Scanner," webpage, undated. As of June 10, 2024:
https://fastna.myshopify.com

SecureDNA, homepage, undated. As of June 10, 2024:
https://securedna.org

SHIELD, "Device Fingerprinting That Stops Fraud, Builds Trust & Drives Growth," webpage, undated. As of June 10, 2024:
https://shield.com/device-fingerprinting

Smith, Hamilton O., Clyde A. Hutchison III, Cynthia Pfannkoch, and J. Craig Venter, "Generating a Synthetic Genome by Whole Genome Assembly: φX174 Bacteriophage from Synthetic Oligonucleotides," *Proceedings of the National Academy of Sciences*, Vol. 100, No. 26, December 2003.

Socure, "Socure ID+," webpage, undated. As of June 10, 2024:
https://www.socure.com/products/socure-id

Thompson Reuters, "CLEAR Adverse Media," webpage, undated-a. As of June 10, 2024:
https://legal.thomsonreuters.com/en/products/clear-adverse-media

Thompson Reuters, "Denied Party Screening Software," webpage, undated-b. As of June 10, 2024:
https://tax.thomsonreuters.com/en/products/onesource-denied-party-screening

Trulioo, homepage, undated. As of June 10, 2024:
https://www.trulioo.com

Tumpey, Terrence M., Christopher F. Basler, Patricia V. Aguilar, Hui Zeng, Alicia Solórzano, David E. Swayne, Nancy J. Cox, Jacqueline M. Katz, Jeffery K. Taubenberger, Peter Palese, and Adolfo García-Sastre, "Characterization of the Reconstructed 1918 Spanish Influenza Pandemic Virus," *Science*, Vol. 310, No. 5745, October 2005.

Twist Bioscience, "Gene Synthesis with Twist Bioscience," webpage, undated. As of June 10, 2024:
https://www.twistbioscience.com/products/genes?tab=overview

U.S. Code, Title 18, Crimes and Criminal Procedure; Part I, Crimes; Chapter 10, Biological Weapons; Section 175c, Variola Virus, January 3, 2012.

U.S. Department of Health and Human Services, Office of the Secretary, "Screening Framework Guidance for Providers of Synthetic Double-Stranded DNA," *Federal Register*, Vol. 75, No. 197, October 13, 2010.

U.S. Department of Health and Human Services, Office of the Secretary, Assistant Secretary for Preparedness and Response, "Review and Revision of the Screening Framework Guidance for Providers of Synthetic Double-Stranded DNA," *Federal Register*, Vol. 85, No. 166, August 26, 2020.

WatchGuard, homepage, undated. As of June 10, 2024:
https://www.watchguard.com

Wells, Elizabeth W., and Michael T. Parker, "Chimeric Viruses Containing Select Agents: The Biology Behind Their Creation, Attenuation, and Exclusion from Regulation," *Health Security*, Vol. 21, No. 5, September/October 2023a.

Wells, Elizabeth W., and Michael T. Parker, "Regulating Select Agent Chimeras: Defining the Problem(s) Through the Lens of SARS-CoV-1/SARS-CoV-2 Chimeric Viruses," *Health Security*, Vol. 21, No. 5, September/October 2023b.

Wheeler, Nicole E., Sarah R. Carter, Tessa Alexanian, Christopher Isaac, Jaime Yassif, and Piers Millet, "Developing a Common Global Baseline for Nucleic Acid Synthesis Screening," *Applied Biosafety*, 2024.

White House, *National Biodefense Strategy and Implementation Plan for Countering Biological Threats, Enhancing Pandemic Preparedness, and Achieving Global Health Security*, October 2022.

WHO—*See* World Health Organization.

Williams, Bridget, and Rowan Kane, *Preventing the Misuse of DNA Synthesis: Five Policy Recommendations to Curb Downside Risk*, Institute For Progress, February 14, 2023.

World Economic Forum and Nuclear Threat Initiative, *Biosecurity Innovation and Risk Reduction: A Global Framework for Accessible, Safe and Secure DNA Synthesis*, January 2020.

World Health Organization, *WHO Advisory Committee on Variola Virus Research: Report of the Eighteenth Meeting*, 2016a.

World Health Organization, "WHO Recommendations Concerning the Distribution, Handling and Synthesis of Variola Virus DNA," revised January 13, 2016b.

World Health Organization, *Global Guidance Framework for the Responsible Use of the Life Sciences: Mitigating Biorisks and Governing Dual-Use Research*, 2022.

Xie, Xuping, Kumari G. Lokugamage, Xianwen Zhang, Michelle N. Vu, Antonio E. Muruato, Vineet D. Menachery, and Pei-Yong Shi, "Engineering SARS-CoV-2 Using a Reverse Genetic System," *Nature Protocols*, Vol. 16, No. 3, March 2021.

Yu, Chuer, Xuhong Zhang, Yuxuan Duan, Senbo Yan, Zonghui Wang, Yang Xiang, Shouling Ji, and Wenzhi Chen, "Diff-ID: An Explainable Identity Difference Quantification Framework for DeepFake Detection," *IEEE Transactions on Dependable and Secure Computing*, 2024.

Zaidenberg, Nezer Jacob, "Game Console Protection and Breaking It," in Mehdi Khosrow-Pour D. B. A. (ed.), *Encyclopedia of Criminal Activities and the Deep Web*, Vol. 2, IGI Global, 2020.

Zhang, Lisa, and Gigi Kwik Gronvall, "Red Teaming the Biological Sciences for Deliberate Threats," *Terrorism and Political Violence*, Vol. 32, No. 6, 2020.

About the Authors

Forrest W. Crawford is a senior statistician at RAND, an associate professor of biostatistics at Yale University, and the senior scientist at Whitespace Ltd., a geospatial intelligence company. He works with the Meselson Center on reducing risk from biological threats and emerging technologies. He holds a Ph.D. in biomathematics.

Kyle Webster is a physical scientist in the Meselson Center at RAND. He is interested in how systems in such contexts as information security and customer screening in finance can be applied to reduce risks from bioweapons. He holds a Ph.D. in biology, specialized in protein engineering.

Gerald L. Epstein is an adjunct fellow at RAND and has spent more than 40 years inside and outside government researching, analyzing, formulating, and communicating public policy at the intersection of science, technology, and security. He works with the Meselson Center on the governance of biotechnology and the life sciences. He holds a Ph.D. in physics.

Derek Roberts is a technical analyst at RAND. He is focused on advances in biotechnology, pathogen surveillance, pandemic response, gene therapy, climate change mitigation, and community resilience. He holds an M.S. in molecular biology.

Joseph Fair is a senior physical scientist with the Meselson Center at RAND. He is a subject-matter expert in biodefense and biosecurity, emerging infectious diseases, and biological threat reduction. He holds a Ph.D. in molecular virology and an M.S. in tropical medicine/epidemiology.

Sella Nevo is a senior information scientist at RAND and director of the Meselson Center, which is dedicated to reducing risks from biological threats and emerging technologies. He holds a B.S. in computer science and a B.S. in mathematics.